NIOSH

Fire Fighter Fatality Investigation and Prevention Program – Selected Investigation Reports

Department of Health and Human Services
Centers for Disease Control and Prevention
National Institute for Occupational Safety and Health

Disclaimer

This document is in the public domain and may be freely copied or reprinted.

Mention of any company or product does not constitute endorsement by the National Institute for Occupational Safety and Health (NIOSH). In addition, citations to Web sites external to NIOSH do not constitute NIOSH endorsement of the sponsoring organizations or their programs or products. Furthermore, NIOSH is not responsible for the content of these Web sites.
All web addresses referenced in this document were accessible as of the publication date.

Ordering Information

To receive documents or other information about occupational safety and health topics, contact
NIOSH at

1-800-CDC-INFO (1-800-232-4636)
TTY: 1-888-232-6348
E-mail: cdcinfo@cdc.gov
or visit the NIOSH Web site at
www.cdc.gov/niosh

For a monthly update on news at NIOSH,
subscribe to NIOSH eNews by visiting
www.cdc.gov/niosh/eNews

DHHS (NIOSH) Publication No. 2011-122
December 2010

SAFER • HEALTHIER • PEOPLE™

Acknowledgments

The principle contributors to this document were Timothy Merinar, Stacy Wertman, Steve Miles, Tommy Baldwin, Dr. Thomas Hales, Matt Bowyer, Virginia Lutz and Jay Tarley. Kimberly Clough Thomas provided editorial and formatting support.

Selected NIOSH fire fighter fatality investigation reports   November 2010

The NIOSH Fire Fighter Fatality Investigation and Prevention Program conducts investigations of fire fighter line-of-duty deaths and injuries to formulate recommendations for preventing future deaths and injuries. The program does not seek to determine fault or place blame on fire departments or individual fire fighters but to learn from these tragic events and prevent future similar events.

The Program Objectives are:

- Better define the characteristics of line-of-duty deaths among fire fighters

- Develop recommendations for the prevention of deaths and injuries

- Disseminate prevention strategies to the fire service.

The enclosed investigation reports are examples of reports for traumatic injury and cardiovascular disease deaths of fire fighters in the last couple of years.

F2009-10	Career Fire Fighter Dies When Backed Over While Spotting an Apparatus – New Jersey	PG 5
F2009-27	Captain Suffers Fatal Heart Attack While Participating in Fire Department Physical Fitness Program – Mississippi	PG 18
F2009-29	Lieutenant Suffers Fatal Heart Attack During Fire Operations - Pennsylvania	PG 30
F2008-34	Volunteer Fire Fighter Dies While Lost in Residential Structure Fire - Alabama	PG 43
F2009-31	One Fire Fighter Killed and Eight Fire Fighters Injured in a Dumpster Explosion at a Foundry – Wisconsin	PG 71

More information about the NIOSH Fire Fighter Fatality Investigation and Prevention Program and a complete listing of all investigation reports and related publications can be found at the program website:
http://www.cdc.gov/niosh/fire

Death in the line of duty...

A summary of a NIOSH fire fighter fatality investigation November 3, 2009

Career Fire Fighter Dies When Backed Over While Spotting an Apparatus—New Jersey

Incident scene
(Photo courtesy of police department.)

SUMMARY

On January 2, 2009, a 57-year-old male career fire fighter (the victim) was fatally injured when he was backed over while spotting an apparatus on the fire scene. The victim was the acting captain the

---

The National Institute for Occupational Safety and Health (NIOSH), an institute within the Centers for Disease Control and Prevention (CDC), is the federal agency responsible for conducting research and making recommendations for the prevention of work-related injury and illness. In fiscal year 1998, the Congress appropriated funds to NIOSH to conduct a fire fighter initiative. NIOSH initiated the Fire Fighter Fatality Investigation and Prevention Program to examine deaths of fire fighters in the line of duty so that fire departments, fire fighters, fire service organizations, safety experts and researchers could learn from these incidents. The primary goal of these investigations is for NIOSH to make recommendations to prevent similar occurrences. These NIOSH investigations are intended to reduce or prevent future fire fighter deaths and are completely separate from the rulemaking, enforcement and inspection activities of any other federal or state agency. Under its program, NIOSH investigators interview persons with knowledge of the incident and review available records to develop a description of the conditions and circumstances leading to the deaths in order to provide a context for the agency's recommendations. The NIOSH summary of these conditions and circumstances in its reports is not intended as a legal statement of facts. This summary, as well as the conclusions and recommendations made by NIOSH, should not be used for the purpose of litigation or the adjudication of any claim. For further information, visit the program Web site at http://www.cdc.gov/niosh/fire or call toll free at 1-800-CDC-INFO (1-800-232-4643).

Fatality Assessment and Control Evaluation Investigation Report # F2009-10

Career Fire Fighter Dies When Backed Over While Spotting an Apparatus—New Jersey

night of the incident and responded in an engine with a crew of three to a reported working structure fire. While en route, the engine had received a radio message to forward lay and supply water for an elevated master stream. Due to the location of the fire structure and hydrant the crew had to lay the supply line beneath a highway overpass. Upon arrival, the engine chauffeur had to drive around a police cruiser and tow truck in order to position the engine to an available hydrant. The engine then dropped off a fire fighter at the hydrant to prepare a forward lay when the incident commander advised them to do a reverse lay. The victim then exited the engine to guide the chauffeur while he backed the engine around the police cruiser and tow truck. The victim walked down the officer's side of the engine and positioned himself at the rear on the officer's side. The fire fighter positioned himself at the driver's side front bumper. The chauffeur was able to negotiate the engine around the police cruiser and tow truck without incident before straightening up to position a feeder line into the scene. The victim walked backwards keeping eye contact with the chauffeur via the officer's side mirror. While backing, the chauffeur noticed the tow truck drive past him toward the scene. He focused his attention on the tow truck momentarily when he felt the truck run over something. A police officer yelled to the chauffeur to stop the engine because something or someone was just run over. The victim was found underneath the engine just in front of the officer's side rear wheels. He was transported to a local metropolitan hospital where he was pronounced dead. The chauffeur was not cited in the fatal incident. Key contributing factors identified in this investigation include loss of direct communications between driver and spotter, driver distractions, possible loss of footing by the victim, and possible failure of the automatic reverse braking system.

NIOSH investigators concluded that, to minimize the risk of similar occurrences, fire departments should

- *Ensure that standard operating procedures (SOPs) are developed, implemented, and enforced on safe backing of fire apparatus (e.g., visual and audio communication, use and position of spotter(s)) and include adequate training and testing methods (e.g. written and practical tests) to ensure fire fighter comprehension.*

- *Consider evaluating current safety equipment used on fire apparatus to assist drivers during backing operations and consider supplementary safety equipment (e.g., additional mirrors, automatic sensing devices, and/or video cameras) for further assistance.*

- *Implement proper procedures for inspection, use, and maintenance of safety equipment used to assist in the backing of fire apparatus to ensure the equipment functions properly when needed.*

INTRODUCTION

On January 2, 2009, a 57-year-old male career fire fighter (the victim) was fatally injured when he was backed over while spotting an apparatus on the fire scene. On March 30, 2009, the International

Fatality Assessment and Control Evaluation
Investigation Report # F2009-10

Career Fire Fighter Dies When Backed Over While Spotting an Apparatus—New Jersey

Association of Fire Fighters requested that the National Institute for Occupational Safety and Health (NIOSH) investigate this incident. On April 13–16, 2009, two safety and occupational health specialists from the NIOSH Fire Fighter Fatality Investigation and Prevention Program traveled to New Jersey to investigate this incident. The NIOSH investigators interviewed the fire chief and fire director from the victim's department, the engine chauffeur and fire fighter, and the fire scene incident commander. The investigators met with the New Jersey Public Employees Occupational Safety and Health Program (PEOSH) incident investigator and representatives and reviewed their photographs, investigative findings, and safety and health standards for fire department personnel. NIOSH investigators also met with representatives from the local fire fighter's union and uniformed officer's union.

Investigators reviewed law enforcement investigative photographs and investigative reports, the New Jersey State Police Commercial Vehicle Crash Report, the autopsy report, training records of the victim and engine chauffeur, and visited the incident scene. NIOSH investigators also spoke with investigators from the New Jersey Division of Fire Safety, local police and sheriff's office, and personnel from the medical examiner's office.

FIRE DEPARTMENT

The career department involved in this incident is comprised of 269 uniformed fire fighters. The department has seven stations and serves a population of approximately 124,000 in a geographical area of 12 square miles.

Although the fire department has provided annual driver/operator training courses that focused on several driving topics including backing, they had not developed and implemented standard operating procedures that will assist fire fighters in safely backing an apparatus. In March 1994, due to an increase in overhead door damage, a general order was issued that addressed procedures for leaving and returning from quarters.[1] In regards to backing, the general order states, "To ensure pedestrian safety and to stop the flow of traffic when returning into quarters, the apparatus shall be guided on both sides." NIOSH investigators observed this general order being used while meeting with the fire department. The State of New Jersey motor vehicle regulations do not require that emergency vehicle operators possess any special training or driver's licenses such as a commercial driver's license.

TRAINING and EXPERIENCE

The victim had been with this department for more than 28 years. He had completed certification courses in Fire Fighter I, Hazardous Materials Operations, Basic First Aid and Cardiopulmonary Resuscitation (CPR), and various weekly and monthly fire service-related topics. The victim had also completed various online and instructor-led training courses on the incident command system.

Fatality Assessment and Control Evaluation Investigation Report # F2009-10

Career Fire Fighter Dies When Backed Over While Spotting an Apparatus—New Jersey

The engine chauffeur had been with this department for more than 18 years. He had completed certification courses in Fire Fighter II, Hazardous Materials Operations, Basic First Aid and CPR, and various weekly and monthly topics related to the fire service. The chauffeur had completed various online and instructor-led training courses on the incident command system.

The victim and engine chauffeur had completed certification training on highway incident and traffic safety, a department-instructed fire apparatus driver/operator course, and National Safety Council's course, *Coaching the Emergency Vehicle Operator II–Fire*.[2] Both apparatus driver/operator courses focused on basic fundamental driving skills such as backing, turning, intersection safety, braking, and speed control. Written and maneuvering skills tests were administered. The engine was equipped with an automatic reverse braking system when the vendor delivered it to the fire department, and the engine chauffeur had received initial training on how the system functioned from the vendor.

EQUIPMENT and PERSONNEL

The victim was the acting officer on the engine along with the chauffeur and fire fighter. The apparatus involved in the incident was a 1996 enclosed cab engine with an automatic transmission, diesel engine, and an air brake system (see Photo 1). The apparatus' gross vehicle weight rating (GVWR) was 32,360 lbs. The engine had two axles with six wheels (two in the front and four in the rear). The engine measured 28 ft (length) x 8 ft 7 in (width) x 9 ft 5 in (height). The apparatus was also equipped with an automatic reverse braking system[3] mounted to the rear step of the apparatus (see Photo 2).

Photo 1. Apparatus involved in the backing incident.
(Photo courtesy of police department.)

Fatality Assessment and Control Evaluation Investigation Report # F2009-10

Career Fire Fighter Dies When Backed Over While Spotting an Apparatus—New Jersey

Photo 2. This reverse braking system is attached to the rear bumper of the apparatus and is designed to actuate the rear brakes when pressure is applied to the rubber sensor.
(Photo courtesy of sheriff's office.)

This automatic reverse braking system was developed in the early 1980s, introduced in the United Kingdom in 1983, and then became available in the United States in 1986. This reverse braking system was an optional safety feature that was installed on the apparatus prior to the fire department taking possession in 1996. The system aids driver/operators while backing an apparatus by locking the rear brakes when the system comes in contact with an object. *Note: The manufacturer's manual notes that this system does not exempt the driver/operator from using conventional backing aids such as side and rear mirrors or spotters.*[3] The rubber sensor is mounted six inches beyond the vehicle's rear end to provide an area where vehicle damage could be reduced due to the short distance traveled after automatic brake application. The sensor has an air chamber which is sealed at either end. The air chamber detects or senses a compression of the wall of the chamber, as little as 0.05 psi, causing the pressure inside to increase and then transmit down to a control/switch unit which triggers the brakes to actuate. The system is designed to operate when the apparatus is in reverse gear and backing at a slow rate of speed. The instruction manual recommends a function test be done after initial installation and at least once a week thereafter. For the involved apparatus, the last documented function test was performed in March 2008; at that time, the system was working properly.

Fatality Assessment and Control Evaluation
Investigation Report # F2009-10

Career Fire Fighter Dies When Backed Over While Spotting an Apparatus—New Jersey

After the incident, the apparatus and scene were processed by local police before the apparatus was escorted by state police to the fire department's headquarters. The apparatus was secured there until state police could perform a vehicle inspection. The state police performed a standard commercial vehicle crash inspection in the presence of a fire department mechanic and investigators from the New Jersey Division of Fire Safety. Items inspected included the air brake system, brakes, steering, suspension, tires, lighting, and low air pressure indicators. The inspection was completed within three days of the incident and the apparatus was found to have no mechanical violations. The state police inspection did not include an inspection of the automatic reverse braking system because it was an optional device and their investigators were not certified to test it. However, when pressure was applied to the rubber sensor while backing, the system failed to actuate the brakes. It could not be determined whether the backing system was functional at the time of the incident, but there was damage to the bumper and rubber sensor that could not be attributed to the fatal incident (see Photos 3 and 4).

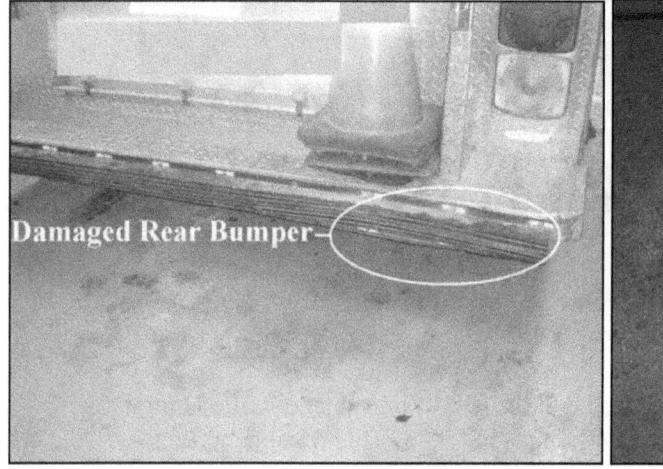

Photos 3 and 4. Pictures show damage to rear bumper and rubber sensor.
(Photo courtesy of PEOSH.)

From these findings, the New Jersey Division of Fire Safety recommended that repairs be made before placing the apparatus back in service. The automatic reverse braking system control unit was replaced following the inspection, and then placed back in service. The fire department also inspected eight other apparatus that had the same reverse braking system installed. Three of these apparatus had damaged or insufficient operating parts within the system that affected proper operation. These three systems were also repaired.

Fatality Assessment and Control Evaluation Investigation Report # F2009-10

Career Fire Fighter Dies When Backed Over While Spotting an Apparatus—New Jersey

TIMELINE

The timeline for this incident includes the initial call to the 911 center at 0213 hours for a structure fire. Only the engine directly involved in the incident is discussed in this timeline. No en route times were documented for initial responding units for the structure fire. The engine was part of the initial assignment dispatched. The engine's dispatch, arrival, and key events include the following:

- **0213 Hours**
  911 dispatch center received a call for a residential structure fire

- **0215 Hours**
  911 center dispatched the residential structure fire assignment

- **0226 Hours**
  Engine on scene

- **0234 Hours**
  Emergency Medical Services transported victim to hospital

PERSONAL PROTECTIVE EQUIPMENT

It was reported to NIOSH investigators that the victim was seen wearing a full array of personal protective clothing and equipment, consisting of turnout gear (coat and pants), helmet, rubber fire boots, and a self-contained breathing apparatus. NIOSH investigators were unable to inspect the victim's gear because it was destroyed prior to this investigation due to biological contamination. The turnout coat and pants were black with reflective trim and lettering. The turnout coat had 3 in reflective lime yellow triple trim stripes around the waist and chest of the coat, and at wrist and elbow locations on both coat arms. There were also (6) 3 in reflective letters on the back of the turnout coat. The turnout pants had 3 in yellow/silver triple trim stripes above the pant cuffs.

WEATHER and ROAD CONDITIONS

The weather at the time of the incident was clear with temperatures below freezing. The incident occurred on a municipal roadway underneath a turnpike overpass approximately 186 ft from the fire scene. The road surface was blacktop and concrete and was straight, level, and dry. No construction occurred in the area of the incident and no concrete dividers existed to separate the eastbound and westbound lanes. The area was reported to have been well lit when the incident occurred.

Fatality Assessment and Control Evaluation
Investigation Report # F2009-10

Career Fire Fighter Dies When Backed Over While Spotting an Apparatus—New Jersey

INVESTIGATION

On January 2, 2009, at 0213 hours, the 911 center received an emergency call for a residential structure fire. At 0215 hours, the 911 center dispatched a residential structure fire assignment which included three engines, a ladder truck, a rescue truck, and a battalion chief. The engine involved in this incident was one of the three initial engines dispatched. The fire scene incident commander initially assigned the engine to forward lay a feeder line to supply water to the ladder truck for an elevated master stream. Due to the layout of the incident scene, the engine had to lay the supply line underneath a highway overpass. The engine approached the fire scene from the west and was positioned at a hydrant approximately 300 feet from the fire scene on the other side of the highway overpass. The engine chauffeur had to negotiate around a police cruiser and a tow truck that were blocking the roadway. The engine's fire fighter exited from the officer's side to prepare the hydrant and pull the feeder line for a forward lay. The fire scene incident commander then radioed the engine and advised them to perform a reverse lay from the aerial truck so that the engine could pump from the hydrant. The victim then exited from the officer's seat, walked down the officer's side and positioned himself at the rear of the engine, officer's side. The victim guided the engine chauffeur as he negotiated around the police cruiser and tow truck in reverse, repositioning the engine so that it was facing away from the fire scene. The fire fighter left the hydrant and then walked over to the driver's side front bumper and positioned himself there to assist the chauffeur in backing as well. Once the engine was straightened up and backing toward the fire scene, the tow truck repositioned on the driver's side. The window, which was down on the driver's side, allowed the chauffeur to briefly look over his left shoulder and observe what the tow truck operator was doing. *Note: The chauffeur stated that he had maintained eye contact with the victim via the officer's side mirror until this occurred.* At the same time the tow truck was being repositioned, the officer of the aerial truck observed the victim walking behind the engine at an angle toward the driver's side and looking toward the fire scene. *Note: It is believed that the victim was trying to get to a better vantage point to assist the chauffeur in backing the engine and may not have realized how close he was to the engine.* The officer of the ladder truck witnessed what appeared to be the victim being hit by the engine and then stumbling. As the engine continued to back up, the victim was run over by the engine.

Witnesses interviewed reported seeing the victim fall down before being backed over by the engine. *Note: The victim may have been contacted by the engine's rear step, or he may have stumbled causing him to fall, before being run over.* The chauffeur remembers looking in his driver's side mirror and seeing a police officer waving his arms in the air and shouting, "Stop, you ran over something or someone!" The chauffeur immediately stopped the engine and was directed by police to shut the engine down, apply the brake, and stay in the engine's cab. The fire fighter grabbed the wheel chocks and chocked the engine. The victim was discovered underneath the engine on the officer's side, just in front of the rear wheels. He was removed from underneath the engine by fire fighters and police officers and then transported to a local metropolitan hospital where he was pronounced dead. Witnesses also noted that the back-up alarm and lights were operating when the incident occurred.

Fatality Assessment and Control Evaluation
Investigation Report # F2009-10

Career Fire Fighter Dies When Backed Over While Spotting an Apparatus—New Jersey

CONTRIBUTING FACTORS

Occupational injuries and fatalities are often the result of one or more contributing factors or key events in a larger sequence of events that ultimately result in the injury or fatality. NIOSH investigators identified the following items as key contributing factors in this incident that ultimately led to the fatality:

- Loss of direct communication between the driver and victim.
- Driver distractions at a crucial time in the incident.
- Possible loss of footing by the victim.
- Possible failure of the automatic reverse braking system to actuate the rear brakes.

CAUSE OF DEATH

According to the county medical examiner's autopsy report, the victim died from blunt force trauma to the head, torso, and upper extremities.

RECOMMENDATIONS

Recommendation #1: Fire departments should ensure that standard operating procedures (SOPs) are developed, implemented, and enforced on safe backing of fire apparatus (e.g., visual and audio communication, use and position of spotter(s)) and include adequate training and testing methods (e.g. written and practical tests) to ensure fire fighter comprehension.

Discussion: The National Fire Protection Association (NFPA) 1500 *Standard on Fire Department Occupational Safety and Health Program* states, "The fire department shall develop standard operating procedures for safely driving fire apparatus during nonemergency travel and emergency response and shall include specific criteria for vehicle speed, crossing intersections, traversing railroad grade crossings, the use of emergency warning devices, and backing of fire apparatus."[4] A SOP on backing fire apparatus should include driver responsibilities (i.e., mirror adjustment and safe path of travel) and the use of a spotter(s). At least one crew member, preferably two crew members, should be positioned to assist a driver during backing operations. A spotter should be positioned at the rear of the fire apparatus on either the driver's or officer's side so that they are visible in the side-view mirrors. If more than one spotter is available, one can be positioned at the rear of the fire apparatus and one at the front or only at the rear, on either side of apparatus. The use of more than one spotter will assist the driver in negotiating tight spaces such as alleyways. The SOP should state that members assigned to assist in backing apparatus be in communication with the driver/operator through the use of department-approved hand signals, one-on-one communication, intercom system, or two-way radio devices. To avoid confusion, it is important to designate only one spotter to communicate with the driver.

Fatality Assessment and Control Evaluation Investigation Report # F2009-10

Career Fire Fighter Dies When Backed Over While Spotting an Apparatus—New Jersey

The National Solid Waste Management Association and the Waste Equipment Technology Association released the *Manual of Recommended Safety Practices*, a comprehensive collection of safety practices for workers engaged in solid waste collection. Refuse truck operators are required to maneuver around vehicles and structures and are faced with having to routinely operate their trucks in reverse on a daily basis. This manual details procedures specific to backing safely, acting as a spotter during backing, and working around mobile equipment.[5] The fire service could adopt these safety practices for backing fire apparatus and use this manual as a reference to develop or evaluate their own backing procedures.

Drivers need to make sure they maintain visual contact with the spotter(s) and stop backing immediately when visual contact is lost. Backing should only be resumed when visual contact is reestablished and the designated spotter gives appropriate direction to continue. It is important that the driver focuses on backing, is not distracted by dispatch traffic, and is not utilizing a cell phone or handheld radio communication device while engaged in backing maneuvers. When spotters are assisting with backing a fire apparatus they need to stay visible within the designated side-view mirrors, stay away from the driver's blind spots, stay clear of the fire apparatus' path of travel, avoid walking backwards, and signal the driver to stop when a person or object comes within the apparatus' path of travel or when the spotter needs to change positions.[6] OSHA regulations require employers to train workers to recognize and avoid unsafe conditions that may be present in their work environments and to provide training on the regulations applicable to their work.[7]

The United States Fire Administration's (USFA) manual, *Safe Operation of Fire Tankers* states, "One of the most common types of crashes that can be traced to failure to follow SOPs is those that involve backing the apparatus."[8] During this incident, the fire apparatus was backing at the same time the victim was observed walking behind the apparatus at an angle. The chauffeur was momentarily distracted by the movement of the tow truck, when this occurred. A general order on backing into a fire station had been issued in 1994[1], but a comprehensive SOP had not been developed or implemented on backing.

Recommendation #2: Fire departments should consider evaluating current safety equipment used on fire apparatus to assist drivers during backing operations and consider supplementary safety equipment (e.g., additional mirrors, automatic sensing devices, and/or video cameras) for further assistance.

Discussion: Modern technology has provided the fire service with a variety of electronic devices that can assist in improving fire apparatus and fire fighter safety. Fire apparatus are currently being equipped with additional mirrors, cameras, and apparatus-mounted sensing devices (e.g., infrared and ultrasonic) to aid in backing maneuvers. Additional mirrors mounted and angled so that blind spots are eliminated can assist the driver while backing. A rear-view camera mounted on the rear of the apparatus provides a view of the obstructed area (blind area directly behind the vehicle) on a video monitor in the cab. Sensor systems (radar or sonar systems designed as backing aids) provide an alarm in the cab when an individual or other obstacle is detected at the rear of an apparatus. A

Fatality Assessment and Control Evaluation Investigation Report # F2009-10

Career Fire Fighter Dies When Backed Over While Spotting an Apparatus—New Jersey

combination of a camera and a sensor system may offer the best protection, especially on a congested fireground or at a motor vehicle incident.

As in this incident, newer apparatus are being equipped with automatic sensing devices, often referred to as Backstops®, that will cause the vehicle's brakes to lock up and stop the apparatus when the device senses contact with an object.[8] This type of device does not prevent the crash but only minimizes potential damage from striking an object. Even though these technological devices may provide an additional measure of safety, they do not substitute for visible spotters.

Recommendation #3: Fire departments should implement proper procedures for inspection, use, and maintenance of safety equipment used to assist in the backing of fire apparatus to ensure the equipment functions properly when needed.

Discussion: NFPA 1500 *Standard on Fire Department Occupational Safety and Health Program* states, "All fire apparatus shall be inspected at least weekly, within 24 hours after any use or repair, and prior to being placed in service or used for emergency purposes, in order to identify and correct unsafe conditions."[4] Written policies and procedures should incorporate a preventative maintenance program that will provide information on proper inspection, maintenance, and repair of apparatus and its equipment. This program should also provide guidance for documenting, notifying, filing, and securing maintenance checks and records, including requirements to place an apparatus out of service. Also, all operating and maintenance instructions and manuals shall be provided and maintained for those performing routine tests, inspections, and servicing functions.[9]

The apparatus involved in this incident was equipped with an after-market reverse braking system which was mounted to the rear bumper. The last documented service test was performed more than eight months prior to the fatal incident. The manufacturer recommends the system be checked daily or weekly to insure proper operation.[3] The system was found to be not working when examined by New Jersey Fire Safety Division investigators during the state police commercial vehicle inspection. However, the system was not tested by the manufacturer or a third-party certification facility to verify that the device failed to actuate the rear brakes. A component for this reverse backing system needed to be replaced to allow it to operate again. *Note: It has not been determined beyond a reasonable doubt that the system failed to operate during the fatal incident or that the victim came into contact with the sensor, but these cannot be ruled out as potential contributing factors in the fatal incident.* Three other apparatus within the fire department's fleet also needed maintenance performed on their installed systems.

Fatality Assessment and Control Evaluation Investigation Report # F2009-10

Career Fire Fighter Dies When Backed Over While Spotting an Apparatus—New Jersey

REFERENCES

1. Fire Department [1994]. General order.

2. NSC [2008]. Coaching the emergency vehicle operator II—fire. Itasca, IL: National Safety Council.

3. Universal Life Safety Products [1996]. Manufacturer manual: automatic reverse braking system. Verona, NJ: Universal Life Safety Products, LLC.

4. NFPA [2007]. NFPA 1500 Standard on fire department occupational safety and health program. 2007 ed. Quincy, MA: National Fire Protection Association.

5. NSWMA [1997]. Manual of recommended safety practices. Washington, DC: National Solid Waste Management Association.

6. NIOSH [1994]. NIOSH Alert: preventing worker injuries and deaths from moving refuse collection vehicles. Cincinnati, OH: U.S. Department of Health and Human Services, Centers for Disease Control and Prevention, National Institute for Occupational Safety and Health, DHHS (NIOSH) Publication No. 97-110 [http://www.cdc.gov/Niosh/refuse.html].

7. Code of Federal Regulations [2009]. 29 CFR 1926.21(b)(2). Safety training and education. Washington, DC: U.S. Printing Office, Office of the Federal Register.

8. USFA/FEMA [2003]. Safe operation of fire tankers. Emmitsburg, MD: U.S. Fire Administration, Publication No. FA 248 [http://www.iaff.org/hs/EVSP/USFA%20Safe%20Operation%20of%20Fire%20Tankers.pdf].

9. NFPA [2007]. NFPA 1451 Standard for a fire service vehicle operations training program. 2007 ed. Quincy, MA: National Fire Protection Association.

INVESTIGATOR INFORMATION

This investigation was conducted by Stacy C. Wertman and Stephen T. Miles, Safety and Occupational Health Specialists, with the Fire Fighter Fatality Investigation and Prevention Program, Fatality Investigations Team, Surveillance and Field Investigations Branch, Division of Safety Research, NIOSH located in Morgantown, WV. This report was authored by Stacy C. Wertman. A technical review was provided by William Peters, of Peters Associates, Fire Apparatus Consulting Services, Inc. Mr. Peters is a retired Battalion Chief with over 28 years experience with the Jersey City, New Jersey Fire Department and is active in the NFPA consensus standards process.

 Fatality Assessment and Control Evaluation
Investigation Report # F2009-10

Career Fire Fighter Dies When Backed Over While Spotting an Apparatus—New Jersey

DISCLAIMER

Mention of any company or product does not constitute endorsement by the National Institute for Occupational Safety and Health (NIOSH). In addition, citations to Web sites external to NIOSH do not constitute NIOSH endorsement of the sponsoring organizations or their programs or products. Furthermore, NIOSH is not responsible for the content of these Web sites.

Death in the line of duty...

A summary of a NIOSH fire fighter fatality investigation — May 2010

Captain Suffers Fatal Heart Attack While Participating in Fire Department Physical Fitness Program – Mississippi

SUMMARY

On October 21, 2009, a 54-year-old male career Captain was jogging alone on a paved trail behind his fire station as a component of the Fire Department's (FD) physical fitness program. The Captain was last seen by crew members at approximately 1630 hours. Approximately 35 minutes later, a civilian found the Captain lying on the trail, unresponsive, and not breathing. The civilian called 911, and a police officer was dispatched. Crew members heard the dispatch and responded to the trail. They found the Captain unresponsive, without a pulse, and not breathing. Crew members began cardiopulmonary resuscitation (CPR) while requesting an ambulance. Despite CPR and advanced life support administered on scene, en route to the hospital's emergency department (ED), and in the ED, the Captain died. The autopsy, completed by the medical examiner, listed "severe coronary artery atherosclerosis with acute plaque change: hemorrhage into an atherosclerotic plaque and rupture of the plaque" as the cause of death. Given the Captain's underlying heart disease, NIOSH investigators conclude that the physical stress of jogging probably triggered a fatal heart attack due to the acute plaque rupture of his right coronary artery.

The NIOSH investigators offer the following recommendations to address general safety and health issues. Had the first recommended measure been in place prior to the Captain's collapse, perhaps his underlying coronary artery disease would have been identified, he would have been referred for treatment, and his sudden cardiac death may have been prevented. The third recommendation may have hastened emergency treatment after the Captain's collapse.

- Ensure fire fighters over the age of 45 with two or more risk factors for CAD have a maximal (symptom-limiting) exercise stress test (EST).

- Ensure fire fighters are cleared for return to duty by a physician knowledgeable about the physical demands of fire fighting, the personal protective equipment used by fire fighters, and the various components of NFPA 1582.

- Ensure on-duty fire fighters exercise in pairs or within viewing distance of another crew member.

Fatality Assessment and Control Evaluation
Investigation Report • F2009–27

Captain Suffers Fatal Heart Attack While Participating in Fire Department Physical Fitness Program – Mississippi

INTRODUCTION & METHODS

On October 21, 2009, a 54-year-old male career Captain suffered sudden cardiac death while exercising during his shift. Despite CPR and advanced life support administered by crew members, the ambulance paramedics, and personnel in the hospital's ED, the Captain died. The United States Fire Administration notified NIOSH of this fatality on October 22, 2009. NIOSH contacted the affected FD to gather additional information on October 26, 2009, and on November 10, 2009, to initiate the investigation. On November 16, 2009, a Safety and Occupational Health Specialist from the NIOSH Fire Fighter Fatality Investigation Team traveled to Mississippi to conduct an on-site investigation of the incident.

During the investigation, NIOSH personnel interviewed the following people:

- Fire Chief
- Training Officer
- Crew members
- The Captain's spouse

NIOSH personnel reviewed the following documents:

- FD policies and operating guidelines
- FD training records
- FD annual report for 2008
- FD incident report
- Police report
- Emergency medical service (ambulance) incident report
- Hospital ED records
- Death certificate
- Autopsy report
- Primary care provider medical records

RESULTS OF INVESTIGATION

Incident. On October 20, 2009, the Captain arrived for duty at Station 3 at approximately 1700 hours for his 24-hour shift. No emergency calls came in during the Captain's shift. At approximately 1630 hours, the Captain left Station 3 to jog on a paved trail in a public park approximately 200 feet behind the fire station. The Captain had been jogging on-duty since the FD began its physical fitness program 12 years ago. He typically jogged at a pace of 5–6 miles per hour (mph) (10–12 minutes to run 1 mile). The temperature was 71 degrees Fahrenheit with 71% relative humidity [NOAA 2009], and the Captain was wearing tennis shoes and an FD-issued T-shirt and shorts. The paved trail covered a little more than one quarter of a mile with some slight uphill and downhill areas.

At approximately 1703 hours, 33 minutes after the Captain left the station, a civilian walking in the park noticed a person lying on the trail. She found him unresponsive and not breathing. She called 911, and a police officer was dispatched. The FD Safety Officer heard the police dispatch and telephoned Station 3 to

Fatality Assessment and Control Evaluation
Investigation Report • F2009–27

Captain Suffers Fatal Heart Attack While Participating in Fire Department Physical Fitness Program – Mississippi

alert them of the call; the Captain's crew members (Engine 3) responded to the area, arriving just ahead of the police.

Crew members found the Captain unresponsive, not breathing, and without a pulse. Dispatch was notified, and an ambulance and the FD (Rescue 1, Deputy Chief, and Fire Chief) were dispatched at 1719 hours. Crew members began CPR and retrieved an automated external defibrillator (AED) and oxygen equipment. The AED advised not to shock. Rescue 1 paramedics arrived at 1724 hours and began advanced life support including cardiac monitoring, intravenous (IV) line placement, and intubation. Proper intubation placement was confirmed by breath sounds and verified by capnography [AHA 2000]. The cardiac monitor revealed asystole (no heart beat), and IV medications were administered. The ambulance arrived on the scene at 1729 hours, and advanced life support treatment continued. The Captain was moved to the ambulance, which departed at 1739 hours en route to the hospital's ED.

The ambulance arrived at the hospital's ED at 1751 hours, at least 45 minutes after his collapse. Inside the ED, resuscitation efforts continued without change in the Captain's clinical condition. At 1800 hours, the Captain was pronounced dead by the attending physician, and resuscitation efforts were stopped.

Medical Findings. The death certificate and the autopsy report, both completed by the medical examiner, listed "complication of coronary artery atherosclerosis" as the cause of death. Autopsy findings showed severe narrowing of all three major coronary arteries, and an acute ruptured plaque completely occluding the right coronary artery. More complete autopsy findings are listed in Appendix A.

The Captain had a history of high blood cholesterol although the exact date of diagnosis was not available to NIOSH investigators. He was prescribed a statin, a cholesterol-lowering medication, in 2008; his cholesterol levels in October 2009 remained elevated. The Captain also had a history of high blood pressure (hypertension). The exact date of diagnosis was not available to NIOSH investigators. He was prescribed an antihypertensive medication, and his blood pressure reading was slightly elevated in October 2009.

As part of the FD annual medical evaluation, the Captain had a "Graded Exercise Stress Test (GXT) - Bicycle Ergometer" test in 2008 and 2009. (These tests had been performed since 1994.) The FD contractor performing this test had subjects pedal on the bicycle at 50 revolutions per minute. Pedal resistance was increased by 0.5 kilogram (kg) each minute until the subjects reached 85% of their maximal age-predicted heart rate, at which point the test was stopped. The contractor used a 12-lead electrocardiogram (EKG) to monitor the subjects' heart rate, and once the test was completed, to look for signs of cardiac ischemia. If ischemia was found, subjects were referred to their primary care physician for follow-up. In addition to pulse, blood pressure was measured every minute during the test and for 5 minutes during the test recovery phase.

Fatality Assessment and Control Evaluation
Investigation Report • F2009–27

Captain Suffers Fatal Heart Attack While Participating in Fire Department Physical Fitness Program – Mississippi

In 2008, the Captain exercised for 4 minutes on the GXT Bicycle Ergometer when he reached 84% of his maximum heart rate, and the test was stopped. He had no chest pain, normal blood pressure response, and no ischemic changes on his EKG. Given his weight of 175 pounds, his estimated aerobic capacity (VO_2) was 20.6 milliliters per kilogram per minute (mL/kg/min) or 5.9 metabolic equivalents (METS) [ACSM 2007]. In 2009, the Captain exercised for 5 minutes when he reached 87% of this maximum heart rate, and the test was stopped. Again, he had no chest pain, normal blood pressure response, and no ischemic changes on his EKG. At this time his weight was 167 pounds, and his estimated VO_2 was 24.8 mL/kg/min or 7.1 METS [ACSM 2007]. In 2008 and 2009, the Captain was "medically cleared under 29 CFR 1910.156, 29 CFR 1910.134, 29 CFR 1910.120 regulations, and NFPA 1582 guidelines" [NFPA 2007a].

Prior to this incident the Captain never reported angina; in June 2008, he had episodes of heartburn that were unrelieved by antacid medications. A scope (esophagogastroduodenoscopy) diagnosed nonerosive gastritis and gastroesphogeal reflux disease (GERD). The symptoms eventually resolved with Nexium®, a medication that reduces stomach acid secretions.

DESCRIPTION OF THE FIRE DEPARTMENT

At the time of the NIOSH investigation, the career FD consisted of three fire stations with 49 uniformed personnel. It served a population of 27,000 residents in a geographic area of 43 square miles.

Employment and Training. The FD requires all new fire fighter applicants to be 21 years of age, have a valid State driver's license, pass a background check, pass a drug screen, pass a physical ability entrance test (PAET) (Appendix B), pass a general aptitude test, and complete an oral interview by a panel that includes FD members and a police officer prior to being ranked. The candidate must then pass a pre-placement medical evaluation and a psychological evaluation prior to being hired. New hires are assigned to day shift during the 8–12 week minimum standards training program for fire fighters. Once the member passes this training program, he/she is placed on a regular duty shift of 24 hours on duty/48 hours off duty, from 1700 hours to 1700 hours. The member then attends the 6-week State Fire Academy to be trained to the NFPA 1001 Fire Fighter I and II level. The member receives emergency medical technician (EMT) training at a local community college. The State requires career fire fighter candidates to meet the State Minimum Standards and Certification Board guideline, which is the National Fire Protection Association (NFPA) 1001, *Standard for Fire Fighter Professional Qualifications* [NFPA 2008]. The Captain was certified as a Fire Fighter II, Driver/Operator, EMT, Fire Officer

Fatality Assessment and Control Evaluation
Investigation Report • F2009–27

Captain Suffers Fatal Heart Attack While Participating in Fire Department Physical Fitness Program – Mississippi

II, HazMat Technician, Fire Service Instructor II, Fire Inspector, Technical Rescue Specialist, and Incident Safety Officer. He had 22 years of fire fighting experience.

Preplacement Medical Evaluations. The FD requires a preplacement medical evaluation for all fire fighter candidates regardless of age. This evaluation includes the following components:

- Complete medical history
- Physical examination (including vital signs)
- Complete blood count with lipid panel
- Pulmonary function test
- Audiogram
- Vision screen
- Urinalysis
- Urine drug screen
- Resting EKG
- Chest x-ray (baseline only)

These evaluations are performed by a physician contracted with the City. Once this evaluation is complete, the contracted physician makes a determination regarding medical clearance for wearing a respirator and fire fighting duties and forwards this decision to the City's personnel director and the FD. The Captain had a preplacement medical evaluation in 1987.

Periodic Medical Evaluations. Annual medical evaluations have been required for all members since 1998. This evaluation includes the following components:

- Complete medical history
- Physical examination (including vital signs)
- Complete blood count with lipid panel
- Pulmonary function test
- Audiogram
- Vision screen
- Urinalysis
- Resting EKG
- Submaximal cycle ergometer test (described in detail on page 4)

These evaluations are performed by a mobile medical clinic. Once this evaluation is complete, a physician traveling with the clinic determines medical clearance for respirator use and fire fighting duties, and forwards this decision to the City's personnel director and the FD.

Return to duty clearance is required for duty-related injuries. The member's primary care physician provides the fire fighter clearance for duty based on NFPA 1582, *Standard on Comprehensive Occupational Medical Program for Fire Departments* [NFPA 2007a]. If members are off duty for 48 hours (2 shifts) continuously for an illness, medical clearance from their primary care physician is required.

Fatality Assessment and Control Evaluation
Investigation Report • F2009–27

Captain Suffers Fatal Heart Attack While Participating in Fire Department Physical Fitness Program – Mississippi

Health and Wellness Programs. The FD has a mandatory wellness/fitness program; exercise (strength and aerobic) equipment is available in the fire stations. Health maintenance programs are available from the City. Self-contained breathing apparatus (SCBA) mask fit tests are performed annually. Semiannual physical ability tests are required for all fire fighters. Fire fighters must perform at least 20 sit-ups, 10 push-ups, and either run 1.5 miles within 16.5 minutes (VO2 of 35.7 mg/kg/min or 10 METs) or walk 3 miles within 45 minutes (VO2 of 14.2 mg/kg/min or 4 METs). The Captain's results are as follows:

DATE	SIT-UPS	PUSH-UPS	RUN
11-14-06	35	20	11:07
05-29-07	35	25	10:25
11-27-07	40	20	11:42
05-27-08	35	20	12:00
01-30-09	35	20	14:39
06-02-09	35	20	13:10

DISCUSSION

Atherosclerotic Cardiovascular Disease. In the United States, atherosclerotic coronary artery disease (CAD) is the most common risk factor for cardiac arrest and sudden cardiac death [Meyerburg and Castellanos 2008]. Risk factors for its development include age over 45, male gender, family history of CAD, smoking, high blood cholesterol, high blood pressure, obesity/physical inactivity, and diabetes [AHA 2009]. The Captain had five of these risk factors (age over 45, male gender, family history of CAD, high blood cholesterol, and high blood pressure) and severe CAD on autopsy.

Narrowing of the coronary arteries by atherosclerotic plaques occurs over many years, typically decades [Libby 2008]. However, the growth of these plaques probably occurs in a nonlinear, often abrupt fashion [Shah 1997]. Heart attacks typically occur with the sudden development of complete blockage (occlusion) in one or more coronary arteries that have not developed a collateral blood supply [Fuster et al. 1992]. This sudden blockage is primarily due to blood clots (thromboses) forming on top of atherosclerotic plaques. The Captain had an acute plaque rupture completely occluding his right coronary artery, establishing that he had an acute heart attack.

Epidemiologic studies have found that heavy physical exertion sometimes immediately precedes and triggers the onset of acute heart attacks and sudden cardiac death [Siscovick et al. 1984; Tofler et al. 1992; Mittleman et al. 1993; Willich et al. 1993; Albert et al. 2000]. Heart attacks in fire fighters have been associated with fire suppression and heavy exertion during training (including physical fitness training) [Kales et al. 2003; Kales et al. 2007; NIOSH 2007]. The Captain had run an unknown number of laps around the jogging track. Assuming his typical pace of approximately 10 minutes per mile (the

Fatality Assessment and Control Evaluation
Investigation Report • F2009–27

Captain Suffers Fatal Heart Attack While Participating in Fire Department Physical Fitness Program – Mississippi

pace required for the physical ability test), the Captain would have expended at least 10 METs in the 35 minutes he was unobserved, which is considered heavy physical activity [Ainsworth et al. 1993]. Given the Captain's underlying CAD, the strenuous physical activity probably triggered a heart attack resulting in his sudden cardiac death.

Left Ventricular Hypertrophy. The autopsy revealed left ventricular hypertrophy (LVH). LVH increases the risk for sudden cardiac death [Levy et al. 1990]. Hypertrophy of the heart's left ventricle is a relatively common finding among individuals with long-standing high blood pressure, a heart valve problem, or chronic cardiac ischemia (coronary artery disease) [Siegel 1997]. The Captain's hypertension and chronic cardiac ischemia were most likely responsible for his LVH.

Occupational Medical Standards for Structural Firefighting. To reduce the risk of sudden cardiac arrest or other incapacitating medical conditions among fire fighters, the NFPA developed NFPA 1582, Standard on Comprehensive Occupational Medical Program for Fire Departments [NFPA 2007a]. NFPA 1582 recommends that all fire fighters receive annual medical evaluations. As part of this medical evaluation, fire fighters should receive a submaximal stress EKG test (85% of the fire fighter's maximal heart rate) as a measure of aerobic capacity. For asymptomatic fire fighters over age 45 (55 for women) with two or more risk factors for CAD (e.g., the Captain), NFPA 1582 recommends a cardiology evaluation with a symptom limiting (maximal) imaging stress test. This recommendation is consistent with the recommendation of the American Heart Association and the American College of Cardiology (AHA/ACC) [Gibbons et al. 2002]. According to the records reviewed by the NIOSH investigator, neither a cardiology evaluation nor an imaging maximal stress test was recommended to the Captain by the FD contract clinic or the Captain's primary care physician. Either may have identified his underlying CAD, resulting in further evaluation and treatment and the likely prevention of his sudden cardiac death.

RECOMMENDATIONS

The NIOSH investigator offers the following recommendations to address general safety and health issues. Had the first recommended measure been in place prior to the Captain's collapse, perhaps his underlying coronary artery disease could have been identified, he would have been referred for treatment, and his sudden cardiac death may have been prevented. The third recommendation may have hastened emergency treatment after the Captain's collapse.

Recommendation #1: Ensure fire fighters over the age of 45 with two or more risk factors for CAD have a maximal (symptom-limiting) exercise stress test (EST).

NFPA 1582, the IAFF/IAFC Fire Service Joint Labor Management Wellness/Fitness Initiative, and the ACC/AHA recommend an exercise stress test for male fire fighters older than 45 with two or more CAD risk factors [IAFF,

Fatality Assessment and Control Evaluation
Investigation Report • F2009–27

Captain Suffers Fatal Heart Attack While Participating in Fire Department Physical Fitness Program – Mississippi

IAFC 2008]. Although the Captain had a submaximal cycle ergometer test to measure his aerobic capacity, this test is not the same as the diagnostic exercise stress test recommended by NFPA 1582 or the ACC/AHA [Gibbons et al. 2002; NFPA 2007a]. The FD contract clinic and/or the Captain's primary care physician should have recommended a maximal EST because the Captain was over the age of 45 and had more than two CAD risk factors.

Recommendation #2: Ensure that fire fighters are cleared for return to duty by a physician knowledgeable about the physical demands of fire fighting, the personal protective equipment used by fire fighters, and the various components of NFPA 1582.

Guidance regarding medical evaluations and examinations for structural fire fighters can be found in NFPA 1582 [NFPA 2007a] and in the IAFF/IAFC Fire Service Joint Labor Management Wellness/Fitness Initiative [IAFF, IAFC 2008]. According to these guidelines, the FD should have a physician who is officially responsible for guiding, directing, and advising the members with regard to their health, fitness, and suitability for duty. The FD physician should review job descriptions and essential job tasks required for all FD positions and ranks to understand the physiological and psychological demands of fire fighters and the environmental conditions under which they must perform. The FD physician should also be familiar with the personal protective equipment the fire fighter wears during various types of emergency operations, and the medical guidance contained in NFPA 1582. If the FD does not have a physician on staff or on contract, the fire fighter's personal physician should be advised of these guidelines.

Recommendation #3: Ensure on-duty fire fighters exercise in pairs or within viewing distance of another crew member.

Members should exercise in pairs or at least within viewing distance of another crew member. If a medical emergency occurs, the other crew member can alert EMS or dispatch. Another option would be for exercising members to carry a PASS device and/or portable radio. PASS devices are portable, lightweight units that, when activated, emit a 95-decibel alarm. The device, which can be manually activated, automatically activates if no motion is detected for approximately 30 seconds [NFPA 2007b].

Portable radios have the advantage of allowing affected members to specify the problem and their exact location. The disadvantages are that a radio is a little larger and heavier, and a radio will not automatically alert anyone if the member suddenly collapses. At the time of this report, the FD requires members to carry portable radios that have a panic button.

REFERENCES

ACSM (American College of Sports Medicine) [2007]. Metabolic equations for gross VO2 in metric units. [http://www.acsm.org/AM/Template.cfm?Section=Exam_Information&Template=/CM/HTMLDisplay.cfm&ContentID=13024]. Date accessed: March 2010.

Captain Suffers Fatal Heart Attack While Participating in Fire Department Physical Fitness Program – Mississippi

AHA [2000]. Advanced cardiovascular life support: section 3: adjuncts for oxygenation, ventilation, and airway control. Circ 102(8)(Suppl):I-95–I-104.

AHA [2009]. AHA scientific position, risk factors for coronary artery disease. Dallas, TX: American Heart Association [http://www.americanheart.org/presenter.jhtml?identifier=4726]. Date accessed: March 2010.

Ainsworth BE, Haskell WL, Leon AS, Jacobs DR Jr., Montoye HJ, Sallis JF, Paffenbarger RS Jr. [1993]. Compendium of physical activities: classification of energy costs of human physical activities. Med Sci Sports Exerc 25(1):71–80.

Albert CM, Mittleman MA, Chae CU, Lee IM, Hennekens CH, Manson JE [2000]. Triggering of sudden death from cardiac causes by vigorous exertion. N Engl J Med 343(19):1355–1361.

Fuster V, Badimon L, Badimon JJ, Chesebro JH [1992]. The pathogenesis of coronary artery disease and the acute coronary syndromes. N Engl J Med 326(4):242–250.

Gibbons RJ, Balady GJ, Bricker JT, Chaitman BR, Fletcher GF, Froelicher VF, Mark DB, McCallister BD, Mooss AN, O'Reilly MG, Winters WL Jr. [2002]. ACC/AHA 2002 guideline update for exercise testing: a report of the American College of Cardiology/American Heart Association Task Force on Practice Guidelines (Committee on Exercise Testing). [http://content.onlinejacc.org/cgi/content/short/40/8/1531]. Date accessed: March 2010.

IAFF, IAFC [2008]. The fire service joint labor management wellness/fitness initiative. 3rd ed. Washington, DC: International Association of Fire Fighters, International Association of Fire Chiefs.

Kales SN, Soteriades ES, Christoudias SG, Christiani DC [2003]. Firefighters and on-duty deaths from coronary heart disease: a case control study. Environ health: a global access science source. 2:14. [http://www.ehjournal.net/content/2/1/14]. Date accessed: March 2010.

Kales SN, Soteriades ES, Christophi CA, Christiani DC [2007]. Emergency duties and deaths from heart disease among fire fighters in the United States. N Engl J Med 356(12):1207–1215.

Levy D, Garrison RJ, Savage DD, Kannel WB, Castelli WP [1990]. Prognostic implications of echocardiographically determined left ventricular mass in the Framingham Heart Study. N Engl J Med 323(24):1706–1707.

Libby P [2008]. The pathogenesis, prevention, and treatment of atherosclerosis. In: Fauci AS, Braunwald E, Kasper DL, Hauser SL, Longo DL, Jameson JL, Loscalzo J, eds. Harrison's principles of internal medicine. 17th ed. New York: McGraw-Hill, pp. 1501–1509.

Meyerburg RJ, Castellanos A [2008]. Cardiovascular collapse, cardiac arrest, and sudden cardiac death. In: Fauci AS, Braunwald E, Kasper DL, Hauser SL, Longo DL, Jameson JL, Loscalzo J, eds. Harrison's principles of internal medicine. 17th ed. New York: McGraw-Hill, pp. 1707–1713.

Mittleman MA, Maclure M, Tofler GH, Sherwood JB, Goldberg RJ, Muller JE [1993]. Triggering of acute myocardial infarction by heavy physical exertion. N Engl J Med 329(23):1677–1683.

NFPA [2007a]. Standard on comprehensive occupational medical program for fire departments. Quincy, MA: National Fire Protection Association. NFPA 1582.

NFPA [2007b]. Standard on personal alert safety systems (PASS). Quincy, MA: National Fire Protection Association. NFPA 1982.

NFPA [2008]. Standard for fire fighter professional qualifications. Quincy, MA: National Fire Protection Association. NFPA 1001.

NIOSH [2007]. NIOSH alert: preventing fire fighter fatalities due to heart attacks and other sudden cardiovascular events. Cincinnati, OH: U.S. Department of Health and Human Services, Public Health Service, Centers for Disease Control and Prevention, National Institute for Occupational Safety and Health, DHHS (NIOSH) Publication No. 2007-113.

NOAA [2009]. Quality controlled local climatological data; hourly observations table; Jackson International Airport, Jackson, MS. National Oceanic and Atmospheric Administration. [http://cdo.ncdc.noaa.gov/qclcd/QCLCD?prior=N]. Date accessed: December 2009.

Shah PK [1997]. Plaque disruption and coronary thrombosis: new insight into pathogenesis and prevention. Clin Cardiol 20 (11 Suppl2):II-38–44.

Siegel RJ [1997]. Myocardial hypertrophy. In: Bloom S, ed. Diagnostic criteria for cardiovascular pathology acquired diseases. Philadelphia, PA: Lippencott-Raven, pp. 55–57.

Siscovick DS, Weiss NS, Fletcher RH, Lasky T [1984]. The incidence of primary cardiac arrest during vigorous exercise. N Engl J Med 311(14):874–877.

Tofler GH, Muller JE, Stone PH, Forman S, Solomon RE, Knatterud GL, Braunwald E [1992]. Modifiers of timing and possible triggers of acute myocardial infarction in the Thrombolysis in Myocardial Infarction Phase II (TIMI II) Study Group. J Am Coll Cardiol 20(5):1049–1055.

Willich SN, Lewis M, Lowel H, Arntz HR, Schubert F, Schroder R [1993]. Physical exertion as a trigger of acute myocardial infarction. N Engl J Med 329(23):1684–1690.

INVESTIGATOR INFORMATION

This incident was investigated by the NIOSH Fire Fighter Fatality Investigation and Prevention Program, Cardiovascular Disease Component in Cincinnati, Ohio. Mr. Tommy Baldwin (MS) led the investigation and co-authored the report. Mr. Baldwin is a Safety and Occupational Health Specialist, a National Association of Fire Investigators (NAFI) Certified Fire and Explosion Investigator, an International Fire Service Accreditation Congress (IFSAC) Certified Fire Officer I, and a former Fire Chief and Emergency Medical Technician. Dr. Thomas Hales (MD, MPH) provided medical consultation and co-authored the report. Dr. Hales is a member of the NFPA Technical Committee on Occupational Safety and Heath, and Vice-Chair of the Public Safety Medicine Section of the American College of Occupational and Environmental Medicine (ACOEM).

Fatality Assessment and Control Evaluation
Investigation Report • F2009–27

Captain Suffers Fatal Heart Attack While Participating in Fire Department Physical Fitness Program – Mississippi

Appendix A

Autopsy Findings

- Severe atherosclerotic CAD
    - Total (100%) focal narrowing of the right coronary artery
    - Severe (80%) focal narrowing of the left anterior descending coronary artery
    - Moderate (70%) focal narrowing of the circumflex coronary artery
    - Acute myocardial infarction (heart attack) of the posterior wall due to acute plaque rupture in the right coronary artery
    - Evidence of a separate posterior wall heart attack 10–14 days prior to his death
- Left ventricular hypertrophy (LVH)
    - Left ventricular and interventricular septum walls thickened (2.0 cm, 1.7 cm respectively); normal by autopsy 0.76–0.88 cm [Colucci and Braunwald 1997]; normal by echocardiography 0.6–1.1 cm [Armstrong and Feigenbaum 2001]
- Normal heart size (390 grams)
- No evidence of a pulmonary embolus (blood clot in the lung arteries)
- Negative blood tests for drugs and alcohol
- Final pathologic diagnosis: "severe coronary artery atherosclerosis with acute plaque change: hemorrhage into an atherosclerotic plaque and rupture of the plaque" as the cause of death

REFERENCES

Armstrong WF, Feigenbaum H [2001]. Echocardiography. In: Braunwald E, Zipes DP, Libby P, eds. Heart disease: a text of cardiovascular medicine. 6th ed. Vol. 1. Philadelphia, PA: W.B. Saunders Company, p. 167.

Colucci WS, Braunwald E [1997]. Pathophysiology of heart failure. In: Braunwald, ed. Heart disease. 5th ed. Philadelphia, PA: W.B. Saunders Company, p. 401.

Silver MM, Silver MD [2001]. Examination of the heart and of cardiovascular specimens in surgical pathology. In: Silver MD, Gotlieb AI, Schoen FJ, eds. Cardiovascular pathology. 3rd ed. Philadelphia, PA: Churchill Livingstone, pp. 8–9.

Fatality Assessment and Control Evaluation
Investigation Report • F2009–27

Captain Suffers Fatal Heart Attack While Participating in Fire Department Physical Fitness Program – Mississippi

Appendix B

Physical Ability Entrance Test

The fire fighter candidate must pass the following components within the time specified:

- 35 bent knee sit-ups in less than 2 minutes
- 65-foot ladder climb without stopping
- 50-foot rescue crawl while wearing a 50-pound vest
- Run 1.5 miles in less than 13 minutes

The National Institute for Occupational Safety and Health (NIOSH), an institute within the Centers for Disease Control and Prevention (CDC), is the federal agency responsible for conducting research and making recommendations for the prevention of work-related injury and illness. In fiscal year 1998, the Congress appropriated funds to NIOSH to conduct a fire fighter initiative. NIOSH initiated the Fire Fighter Fatality Investigation and Prevention Program to examine deaths of fire fighters in the line of duty so that fire departments, fire fighters, fire service organizations, safety experts and researchers could learn from these incidents. The primary goal of these investigations is for NIOSH to make recommendations to prevent similar occurrences. These NIOSH investigations are intended to reduce or prevent future fire fighter deaths and are completely separate from the rulemaking, enforcement and inspection activities of any other federal or state agency. Under its program, NIOSH investigators interview persons with knowledge of the incident and review available records to develop a description of the conditions and circumstances leading to the deaths in order to provide a context for the agency's recommendations. The NIOSH summary of these conditions and circumstances in its reports is not intended as a legal statement of facts. This summary, as well as the conclusions and recommendations made by NIOSH, should not be used for the purpose of litigation or the adjudication of any claim. For further information, visit the program website at

www.cdc.gov/niosh/fire/
or call toll free
1–800–CDC–INFO (1–800–232–4636)

Death in the line of duty...

A summary of a NIOSH fire fighter fatality investigation
March 2010

Lieutenant Suffers Fatal Heart Attack During Fire Operations – Pennsylvania

SUMMARY

On October 24, 2009, a 41-year-old male volunteer lieutenant (LT) responded to a reported residential fire with possible entrapment. At the scene, the LT assisted in stretching a 2-inch hoseline and participated in extinguishing the fire. After about 16 minutes, the water supply ran low, and crews took a break. The LT complained of a headache as he climbed into his engine's cab. The on-scene ambulance crew found the LT in the cab sweating heavily, complaining of tightness in his chest and shortness of breath. The LT became semiconscious and was helped to the ground and then carried to the ambulance. The cardiac monitor showed changes diagnostic of a heart attack. While en route to the hospital's emergency department (ED), the LT's condition worsened and, as the ambulance arrived at the ED, the LT suffered cardiac arrest. Cardiopulmonary resuscitation (CPR) and advanced life support were begun and continued in the ED for over an hour until the ED physician pronounced him dead. The death certificate and the autopsy listed "severe atherosclerotic coronary artery disease (CAD) and hypertensive cardiomyopathy" as the cause of death. Given the LT's severe underlying CAD, NIOSH investigators concluded that the physical exertion involved in responding to the call, stretching the fire hose, and extinguishing the fire triggered a heart attack and sudden cardiac death.

NIOSH investigators offer the following recommendations to address general safety and health issues.

- Provide preplacement and annual medical evaluations to all fire fighters.
- Perform a preplacement and an annual physical performance (physical ability) evaluation.
- Ensure fire fighters are cleared for return to duty by a physician knowledgeable about the physical demands of fire fighting, the personal protective equipment used by fire fighters, and the various components of National Fire Protection Association (NFPA) 1582.
- Phase in a comprehensive wellness and fitness program for fire fighters.
- Provide fire fighters with medical clearance to wear self-contained breathing apparatus (SCBA) as part of the Fire Department's medical evaluation program.
- Conduct annual respirator fit testing.

Fatality Assessment and Control Evaluation Investigation Report • F2009-29

Lieutenant Suffers Fatal Heart Attack During Fire Operations – Pennsylvania

INTRODUCTION & METHODS

On October 24, 2009, a 41-year-old male volunteer LT died while fighting a structure fire. NIOSH was notified of this fatality on October 26, 2009, by the U.S. Fire Administration. NIOSH contacted the affected Fire Department (FD) on October 26, 2009, to gather additional information, and on December 1, 2009, to initiate the investigation. On December 7, 2009, a safety and occupational health specialist from the NIOSH Fire Fighter Fatality Investigation Team traveled to Pennsylvania to conduct an on-site investigation of the incident.

During the investigation, NIOSH personnel interviewed the following people:
- Fire Chief
- Crew members
- LT's family

NIOSH personnel reviewed the following documents:
- FD training records
- FD annual report for 2008
- FD incident report
- Emergency medical service (ambulance) incident report
- Hospital ED records
- Death certificate
- Autopsy report
- Primary care provider medical records
- Employer medical records

INVESTIGATIVE RESULTS

Incident. On October 24, 2009, the FD was dispatched at 0013 hours to a residential structure fire with possible entrapment. The LT responded from his home to the fire station. One engine (staffed with a Driver/Operator, the Incident Commander, the LT, and a fire fighter) and an engine/tanker (staffed with three FD personnel) responded, arriving on the scene at 0022 hours. Upon arrival the crew realized that the residence on fire had been abandoned for many years. During size-up, the Incident Commander noted the roof and second floor had already collapsed into the basement and called for a "defensive attack." Three personnel, including the LT, deployed a 2-inch hose to the "D" side of the residence and began exterior fire suppression. Crew members initially wore full turnout gear and SCBA (not on air) but after a short time, they removed their SCBAs.

The Chief arrived at about 0025 hours, assumed command, and ordered a deck gun to replace the the hoseline. During this transition, the "B" side of the building collapsed into the basement. Additional FD personnel and mutual aid responded and arrived throughout the incident including an ambulance staffed with two emergency medical technicians (EMTs) and one paramedic.

After approximately 16 minutes on scene (0038 hours), both engines and the portable dump tank exhausted their water supplies. While the engine/tanker was en route to replenish its water, crew members began taking breaks. The LT informed a crew member that his head hurt

Fatality Assessment and Control Evaluation
Investigation Report • F2009–29

Lieutenant Suffers Fatal Heart Attack During Fire Operations – Pennsylvania

as he climbed into his engine's cab. The crew member asked nearby EMTs to evaluate the LT. The EMTs found the LT sitting in his cab, wearing bunker pants, boots, and a shirt. He was sweating heavily and complaining of lightheadedness with some tightness in his chest. As a stretcher was being retrieved, the LT became limp. The paramedic and nearby fire fighters assisted the LT to the ground where, although his eyes were closed, he was arousable and had a strong radial pulse. The EMT, paramedic, and two fire fighters carried the LT to the stretcher and loaded him into the ambulance.

Oxygen was administered as a cardiac monitor showed a heart rate of 110 beats per minute and significant ST-segment elevation (diagnostic of a heart attack). Two aspirin were given; two attempts at starting an intravenous (IV) line were unsuccessful. The LT became combative, not allowing measurement of his blood pressure or administration of nitroglycerin. However, the EMT was able to confirm a strong, regular radial pulse.

The ambulance departed the scene at 0047 hours en route to the local hospital's emergency department (ED). During the transport, the LT sat upright and attempted to remove the oxygen mask, stating that he was having difficulty breathing. The crew calmed the LT and replaced the mask. He attempted again to remove the oxygen mask, and the paramedic replaced the mask with a nasopharyngeal airway. Upon nearing the hospital, the LT's heart rate dropped to 38 beats per minute, and cardiac pacing was initiated at a rate of 70 beats per minute with capture.

As the ambulance arrived at the ED (0057 hours), the LT became unresponsive, stopped breathing, and was pulseless. The LT was transferred to the ED staff who began CPR, intubated the LT, and placed an IV line. Cardiac resuscitation medications were administered, and the LT was shocked (defibrillated) two times with no change in his heart rhythm. Advanced life support continued without change in the LT's condition until 0206 hours, when the attending physician pronounced the LT dead, and resuscitation efforts stopped.

Medical Findings. The death certificate, completed by the coroner, and the autopsy, completed by the forensic pathologist, listed "severe atherosclerotic coronary artery disease and hypertensive cardiomyopathy" as the cause of death. Specific findings from the autopsy are listed in Appendix A.

The LT was 69 inches tall and weighed 220 pounds, giving him a body mass index (BMI) of 32.5 kilograms per meters squared (kg/m2). A BMI > 30.0 kilograms per meter squared is considered obese [CDC 2010]. The LT's risk factors for coronary artery disease (CAD) included hypertension (high blood pressure), smoking, family history of CAD, and obesity/lack of exercise. He had been prescribed two antihypertensive medications but had not refilled the prescriptions for over a year. The LT had gone hunting the day before his death. He had no medical complaints. He had walked over somewhat hilly terrain for approximately 2 miles, expending about 9 metabolic equivalents (METs), which is considered moderate physical activity [Ainsworth et al. 1993; Peterson et al. 1999].

Fatality Assessment and Control Evaluation
Investigation Report • F2009–29

Lieutenant Suffers Fatal Heart Attack During Fire Operations – Pennsylvania

DESCRIPTION OF THE FIRE DEPARTMENT

At the time of the NIOSH investigation, the volunteer FD consisted of one fire station with 25 uniformed personnel that served 2,000 residents in a geographic area of 30 square miles. In 2008, the FD responded to 55 calls including 17 structure fires, 2 brush fires, 2 vehicle fires, 7 motor vehicle crashes, and 27 other calls.

Membership and Training. The FD requires new fire fighter applicants to be recommended by three currently active FD members, complete an application, be 18 years of age (21 years to drive fire apparatus), and have a valid State driver's license. The applicant is voted on at the next general meeting. The member must complete the Fire Fighter Essentials course to fight interior structure fires. The FD also has a Junior Fire Fighter Program that allows the junior fire fighter to train and perform support activities until reaching the age of 18. Pennsylvania does not require minimum training levels for fire fighters. The LT was certified as a Fire Fighter, Driver/Operator, Wildland Fire Fighter, and in water rescue. He had 28 years of fire fighting experience.

Preplacement and Periodic Medical Evaluations. The FD does not require preplacement or periodic (annual) medical evaluations for members. No annual SCBA medical clearance or annual SCBA facepiece fit test are required. Members injured on duty must be evaluated by their primary care physician, who makes the final determination regarding return to duty.

Health and Wellness Programs. The FD does not have a formal wellness/fitness program, and no exercise equipment is available in the fire station. No annual physical ability test is required.

DISCUSSION

Atherosclerotic Coronary Artery Disease. In the United States, atherosclerotic CAD is the most common risk factor for cardiac arrest and sudden cardiac death [Meyerburg and Castellanos 2008]. Risk factors for its development include age older than 45, male gender, family history of CAD, smoking, high blood pressure, high blood cholesterol, obesity/physical inactivity, and diabetes [AHA 2010; NHLBI 2010]. The LT had five CAD risk factors (male gender, family history of CAD, smoking, hypertension, and obesity/lack of exercise); the autopsy revealed severe CAD.

The narrowing of the coronary arteries by atherosclerotic plaques occurs over many years, typically decades [Libby 2008]. However, the growth of these plaques probably occurs in a nonlinear, often abrupt fashion [Shah 1997]. Heart attacks typically occur with the sudden development of complete blockage (occlusion) in one or more coronary arteries that have not developed a collateral blood supply [Fuster et al. 1992]. This sudden blockage is primarily due to blood clots (thromboses) forming on top of atherosclerotic plaques.

Fatality Assessment and Control Evaluation Investigation Report • F2009–29

Lieutenant Suffers Fatal Heart Attack During Fire Operations – Pennsylvania

Establishing the occurrence of a recent (acute) heart attack requires any of the following: characteristic electrocardiogram (EKG) changes, elevated cardiac enzymes, or coronary artery thrombus. In the LT's case, his symptoms (chest tightness, lightheadedness, shortness of breath after exertion, and sweating) were typical of a heart attack. In addition, the ambulance's cardiac monitor revealed significant ST-segment elevation, a finding that confirms an acute heart attack.

Epidemiologic studies have found that heavy physical exertion sometimes immediately precedes and triggers the onset of acute heart attacks and sudden cardiac death [Siscovick et al. 1984; Tofler et al. 1992; Mittleman et al. 1993; Willich et al. 1993; Albert et al. 2000]. Heart attacks in fire fighters have been associated with alarm response, fire suppression, and heavy exertion during training (including physical fitness training) [Kales et al. 2003; Kales et al. 2007; NIOSH 2007]. The LT had responded to the alarm, stretched a hoseline, and performed fire suppression. These activities expended about 8 METs, which is considered moderate physical activity [AIHA 1971; Gledhill and Jamnik 1992]. Given the LT's underlying CAD, the stress of performing fire fighting duties probably triggered his acute heart attack, resulting in sudden cardiac death.

Cardiomegaly/Left Ventricular Hypertrophy. On autopsy, the LT was found to have left ventricular hypertrophy (LVH) and an enlarged heart. Both LVH and cardiomegaly increase the risk for sudden cardiac death [Levy et al. 1990]. Hypertrophy of the heart's left ventricle is a relatively common finding among individuals with long-standing high blood pressure (hypertension), a heart valve problem, or chronic cardiac ischemia (reduced blood supply to the heart muscle) [Siegel 1997]. The LT had a history of hypertension and CAD consistent with chronic cardiac ischemia.

Occupational Medical Standards for Structural Fire Fighters. To reduce the risk of sudden cardiac arrest or other incapacitating medical conditions among fire fighters, the NFPA developed NFPA 1582, Standard on Comprehensive Occupational Medical Program for Fire Departments [NFPA 2007a]. This voluntary industry standard provides (1) the components of a preplacement and annual medical evaluation and (2) medical fitness for duty criteria. The LT had a history of hypertension but had not taken his prescription antihypertensive medication for over a year.

Exercise stress tests screen people at risk for CAD and sudden cardiac death. NFPA 1582 recommends performing an exercise stress test "as clinically indicated by history or symptoms" and refers the reader to Appendix A [NFPA 2007a]. Items in Appendix A are not standard requirements, but are provided for "informational purposes only." Appendix A recommends using submaximal (85% of predicted heart rate) stress tests as a screening tool to evaluate a fire fighter's aerobic capacity. Diagnostic stress tests (maximal or symptom-limiting stress tests) with imaging should be used for fire fighters with the following conditions:

Fatality Assessment and Control Evaluation
Investigation Report • F2009–29

Lieutenant Suffers Fatal Heart Attack During Fire Operations – Pennsylvania

- abnormal screening submaximal tests

- cardiac symptoms

- known coronary artery disease

- two or more risk factors for CAD (in men older than 45 and women older than 55)

Risk factors are defined as hypercholesterolemia (total cholesterol greater than 240 milligrams per deciliter [mg/dL]), hypertension (diastolic blood pressure greater than 90 millimeters of mercury [mmHg]), smoking, diabetes mellitus, or family history of premature coronary artery disease (heart attack or sudden cardiac death in a first-degree relative less than 60 years old). This exercise stress test recommendation is similar to that recommended by the American College of Cardiology/American Heart Association (ACC/AHA) and the U.S. Department of Transportation [Gibbons et al. 2002; Blumenthal et al. 2007].

Although the LT had three of the five risk factors listed above for conducting an exercise stress test for CAD, he was younger than the guidelines indicate and therefore would not have been tested.

RECOMMENDATIONS

NIOSH investigators offer the following recommendations to address general safety and health issues.

Recommendation #1: Provide preplacement and annual medical evaluations to all fire fighters.

Guidance regarding the content and frequency of these medical evaluations can be found in NFPA 1582 and in the International Association of Fire Fighters (IAFF)/International Association of Fire Chiefs (IAFC) Fire Service Joint Labor Management Wellness/Fitness Initiative [NFPA 2007a; IAFF, IAFC 2008]. These evaluations are performed to determine fire fighters' medical ability to perform duties without presenting a significant risk to the safety and health of themselves or others. However, the FD is not legally required to follow this standard or this initiative. Applying this recommendation involves economic repercussions and may be particularly difficult for small volunteer fire departments to implement.

To overcome the financial obstacle of medical evaluations, the FD could urge current members to get annual medical clearances from their private physicians. Another option is having the annual medical evaluations completed by paramedics and emergency medical technicians (EMTs) from the local EMS (vital signs, height, weight, visual acuity, and EKG). This information could then be provided to a community physician (perhaps volunteering his or her time), who could review the data and provide medical clearance (or further evaluation, if needed). The more extensive portions of the medical evaluations could be performed by a private physician at the fire fighter's expense (personal or through insurance), provided by a physician volunteer, or paid for by the FD, City, or State. Sharing the financial responsibility for these evaluations be-

Lieutenant Suffers Fatal Heart Attack During Fire Operations – Pennsylvania

tween fire fighters, the FD, the City, the State, and physician volunteers may reduce the negative financial impact on recruiting and retaining needed fire fighters.

Recommendation #2: Perform a preplacement and an annual physical performance (physical ability) evaluation.

NFPA 1500, Standard on Fire Department Occupational Safety and Health Program, requires the FD to develop physical performance requirements for candidates and members who engage in emergency operations [NFPA 2007b]. Members who engage in emergency operations must be annually qualified (physical ability test) as meeting these physical performance standards for structural fire fighters [NFPA 2007b].

Recommendation #3: Ensure that fire fighters are cleared for return to duty by a physician knowledgeable about the physical demands of fire fighting, the personal protective equipment used by fire fighters, and the various components of NFPA 1582.

Guidance regarding medical evaluations and examinations for structural fire fighters can be found in NFPA 1582 [NFPA 2007a] and in the IAFF/IAFC Fire Service Joint Labor Management Wellness/Fitness Initiative [IAFF, IAFC 2008]. According to these guidelines, the FD should have an officially designated physician who is responsible for guiding, directing, and advising the members with regard to their health, fitness, and suitability for duty. The physician should review job descriptions and essential job tasks required for all FD positions and ranks to understand the physiological and psychological demands of fire fighters and the environmental conditions under which they must perform, as well as the personal protective equipment they must wear during various types of emergency operations. This recommendation is based on review of the FD health and medical programs. The LT's primary care physician had treated the LT for hypertension and back pain since 2003. Whether the LT's physician was aware that his patient was a fire fighter is unclear, and neither his medical nor his FD records mentioned medical clearance for duty.

Recommendation #4: Phase in a comprehensive wellness and fitness program for fire fighters.

Guidance for fire department wellness/fitness programs to reduce risk factors for cardiovascular disease and improve cardiovascular capacity is found in NFPA 1583, Standard on Health-Related Fitness Programs for Fire Fighters, the IAFF/IAFC Fire Service Joint Labor Management Wellness/Fitness Initiative, and the National Volunteer Fire Council (NVFC) Health and Wellness Guide, and in Firefighter Fitness: A Health and Wellness Guide [USFA 2004; IAFF, IAFC 2008; NFPA 2008; Schneider 2010]. Worksite health promotion programs have been shown to be cost effective by increasing productivity, reducing absenteeism, and reducing the number of work-related injuries and lost work days [Stein et al. 2000; Aldana 2001]. Fire service health promotion programs have been shown to reduce coronary artery disease risk factors and improve fitness levels, with mandatory programs showing the most benefit [Dempsey et al. 2002; Womack et

Fatality Assessment and Control Evaluation Investigation Report • F2009–29

Lieutenant Suffers Fatal Heart Attack During Fire Operations – Pennsylvania

al. 2005; Blevins et al. 2006]. A study conducted by the Oregon Health and Science University reported a savings of more than $1 million for each of four large fire departments implementing the IAFF/IAFC wellness/fitness program compared to four large fire departments not implementing a program. These savings were primarily due to a reduction of occupational injury/illness claims with additional savings expected from reduced future nonoccupational healthcare costs [Kuehl 2007].

Given the FD's structure, the NVFC program might be the most appropriate model [USFA 2004]. NIOSH recommends a formal, structured wellness/fitness program to ensure all members receive the benefits of a health promotion program.

Recommendation #5: Provide fire fighters with medical clearance to wear SCBA as part of the Fire Department's medical evaluation program.

The Occupational Safety and Health Administration (OSHA) Revised Respiratory Protection Standard requires employers to provide medical evaluations and clearance for employees using respiratory protection [29 CFR 1910.134]. These clearance evaluations are required for private industry employees and public employees in States operating OSHA-approved State plans [OSHA 2009]. Pennsylvania does not operate an OSHA-approved State plan nor other State-regulated workplace safety and health program. However, we recommend voluntary compliance with this standard to ensure all members have been medically cleared to wear an SCBA.

Recommendation #6: Conduct annual respirator fit testing.

The OSHA respiratory protection standard requires employers whose employees are required to use a respirator (e.g., an SCBA) to have a formal respiratory protection program, including annual fit testing [29 CFR 1910.134]. As mentioned previously, Pennsylvania is not an OSHA-approved State plan; therefore, the FD is not required to follow OSHA standards [OSHA 2010].

REFERENCES

AHA [2010]. AHA scientific position, risk factors for coronary artery disease. Dallas, TX: American Heart Association. [http://www.americanheart.org/presenter.jhtml?identifier=4726]. Date accessed: March 2010.

AIHA [1971]. Ergonomics guide to assessment of metabolic and cardiac costs of physical work. Am Ind Hyg Assoc J 32(8):560–564.

Ainsworth BE, Haskell WL, Leon AS, Jacobs DR Jr., Montoye HJ, Sallis JF, Paffenbarger RS Jr. [1993]. Compendium of physical activities: classification of energy costs of human physical activities. Med Sci Sports Exerc 25(1):71–80.

Albert CM, Mittleman MA, Chae CU, Lee IM, Hennekens CH, Manson JE [2000]. Triggering of sudden death from cardiac causes by vigorous exertion. N Engl J Med 343(19):1355–1361.

Aldana SG [2001]. Financial impact of health promotion programs: a comprehensive review of the literature. Am J Health Promot 15(5):296–320.

Fatality Assessment and Control Evaluation
Investigation Report • F2009–29

Lieutenant Suffers Fatal Heart Attack During Fire Operations – Pennsylvania

Blevins JS, Bounds R, Armstrong E, Coast JR [2006]. Health and fitness programming for fire fighters: does it produce results? Med Sci Sports Exerc 38(5):S454.

Blumenthal RS, Epstein AE, Kerber RE [2007]. Expert panel recommendations. Cardiovascular disease and commercial motor vehicle driver safety. [http://www.mrb.fmcsa.dot.gov/documents/CVD_Commentary.pdf]. Date accessed: March 2010.

CDC (Centers for Disease Control and Prevention) [2010]. BMI – Body Mass Index. [http://www.cdc.gov/healthyweight/assessing/bmi/]. Date accessed: March 2010.

CFR. Code of Federal Regulations. Washington, DC: U.S. Government Printing Office, Office of the Federal Register.

Dempsey WL, Stevens SR, Snell CR [2002]. Changes in physical performance and medical measures following a mandatory firefighter wellness program. Med Sci Sports Exerc 34(5):S258.

Fuster V, Badimon L, Badimon JJ, Chesebro JH [1992]. The pathogenesis of coronary artery disease and the acute coronary syndromes. N Engl J Med 326(4):242–250.

Gibbons RJ, Balady GJ, Bricker JT, Chaitman BR, Fletcher GF, Froelicher VF, Mark DB, McCallister BD, Mooss AN, O'Reilly MG, Winters WL Jr., Antman EM, Alpert JS, Faxon DP, Fuster V, Gregoratos G, Hiratzka LF, Jacobs AK, Russell RO, Smith SC Jr [2002]. ACC/AHA 2002 guideline update for exercise testing: a report of the American College of Cardiology/American Heart Association Task Force on Practice Guidelines. Circulation 106(14):1883–1892.

Gledhill N, Jamnik VK [1992]. Characterization of the physical demands of firefighting. Can J Spt Sci 17(3):207–213.

IAFF, IAFC [2008]. The fire service joint labor management wellness/fitness initiative. 3rd ed. Washington, DC: International Association of Fire Fighters, International Association of Fire Chiefs.

Kales SN, Soteriades ES, Christoudias SG, Christiani DC [2003]. Firefighters and on-duty deaths from coronary heart disease: a case control study. Environ health: a global access science source. 2:14. [http://www.ehjournal.net/content/2/1/14]. Date accessed: March 2010.

Kales SN, Soteriades ES, Christophi CA, Christiani DC [2007]. Emergency duties and deaths from heart disease among fire fighters in the United States. N Engl J Med 356(12):1207–1215.

Kuehl K [2007]. Economic impact of the wellness fitness initiative. Presentation at the 2007 John P. Redmond Symposium in Chicago, IL on October 23, 2007.

Levy D, Garrison RJ, Savage DD, Kannel WB, Castelli WP [1990]. Prognostic implications of echocardiographically determined left ventricular mass in the Framingham Heart Study. N Engl J Med 323(24):1706–1707.

Fatality Assessment and Control Evaluation Investigation Report • F2009–29

Lieutenant Suffers Fatal Heart Attack During Fire Operations – Pennsylvania

Libby P [2008]. The pathogenesis, prevention, and treatment of atherosclerosis. In: Fauci AS, Braunwald E, Kasper DL, Hauser SL, Longo DL, Jameson JL, Loscalzo J, eds. Harrison's principles of internal medicine. 17th ed. New York: McGraw-Hill, pp. 1501–1509.

Meyerburg RJ, Castellanos A [2008]. Cardiovascular collapse, cardiac arrest, and sudden cardiac death. In: Fauci AS, Braunwald E, Kasper DL, Hauser SL, Longo DL, Jameson JL, Loscalzo J, eds. Harrison's principles of internal medicine. 17th ed. New York: McGraw-Hill, pp. 1707–1713.

Mittleman MA, Maclure M, Tofler GH, Sherwood JB, Goldberg RJ, Muller JE [1993]. Triggering of acute myocardial infarction by heavy physical exertion. N Engl J Med 329(23):1677–1683.

NFPA [2007a]. Standard on comprehensive occupational medical program for fire departments. Quincy, MA: National Fire Protection Association. NFPA 1582.

NFPA [2007b]. Standard on fire department occupational safety and health program. Quincy, MA: National Fire Protection Association. NFPA 1500.

NFPA [2008]. Standard on health-related fitness programs for fire fighters. Quincy, MA: National Fire Protection Association. NFPA 1583.

NHLBI [2010]. Who is at risk for coronary artery disease? National Heart, Lung, and Blood Institute. [http://www.nhlbi.nih.gov/health/dci/Diseases/Cad/CAD_WhoIsAtRisk.html]. Date accessed: March 2010.

NIOSH [2007]. NIOSH alert: preventing fire fighter fatalities due to heart attacks and other sudden cardiovascular events. Cincinnati, OH: U.S. Department of Health and Human Services, Centers for Disease Control and Prevention, National Institute for Occupational Safety and Health, DHHS (NIOSH) Publication No. 2007-113.

OSHA [2010]. State Occupational Safety and Health Plans. [http://www.osha.gov/fso/osp/index.html]. Date accessed: March 2010.

Peterson AT, Steffen J, Terry L, Davis J, Porcari JP, Foster C [1999]. Metabolic responses associated with deer hunting. Med Sci Sports Exerc 31(12):1844–1848.

Schneider EL [2010]. Firefighter fitness: a health and wellness guide. New York, NY: Nova Science Publishers.

Shah PK [1997]. Plaque disruption and coronary thrombosis: new insight into pathogenesis and prevention. Clin Cardiol 20(11 Suppl2):II-38–44.

Siegel RJ [1997]. Myocardial hypertrophy. In: Bloom S, ed. Diagnostic criteria for cardiovascular pathology acquired diseases. Philadelphia, PA: Lippencott-Raven, pp. 55–57.

Siscovick DS, Weiss NS, Fletcher RH, Lasky T [1984]. The incidence of primary cardiac arrest during vigorous exercise. N Engl J Med 311(14):874–877.

Stein AD, Shakour SK, Zuidema RA [2000]. Financial incentives, participation in employer sponsored health promotion, and changes in employee health and productivity: HealthPlus health quotient program. J Occup Environ Med 42(12):1148–1155.

Tofler GH, Muller JE, Stone PH, Forman S, Solomon RE, Knatterud GL, Braunwald E [1992]. Modifiers of timing and possible triggers of acute myocardial infarction in the thrombolysis in myocardial infarction phase II (TIMI II) study group. J Am Coll Cardiol 20(5):1049–1055.

USFA [2004]. Health and wellness guide. Emmitsburg, MD: Federal Emergency Management Agency; United States Fire Administration. Publication No. FA-267.

Willich SN, Lewis M, Lowel H, Arntz HR, Schubert F, Schroder R [1993]. Physical exertion as a trigger of acute myocardial infarction. N Engl J Med 329(23):1684–1690.

Womack JW, Humbarger CD, Green JS, Crouse SF [2005]. Coronary artery disease risk factors in firefighters: effectiveness of a one-year voluntary health and wellness program. Med Sci Sports Exerc 37(5):S385.

INVESTIGATOR INFORMATION

This incident was investigated by the NIOSH Fire Fighter Fatality Investigation and Prevention Program, Cardiovascular Disease Component in Cincinnati, Ohio. Mr. Tommy Baldwin (MS) led the investigation and co-authored the report. Mr. Baldwin is a Safety and Occupational Health Specialist, a National Association of Fire Investigators (NAFI) Certified Fire and Explosion Investigator, an International Fire Service Accreditation Congress (IFSAC) Certified Fire Officer I, and a former Fire Chief and Emergency Medical Technician. Dr. Thomas Hales (MD, MPH) provided medical consultation and co-authored the report. Dr. Hales is a member of the NFPA Technical Committee on Occupational Safety and Heath, and Vice-Chair of the Public Safety Medicine Section of the American College of Occupational and Environmental Medicine (ACOEM).

APPENDIX A - Autopsy Findings

- Severe atherosclerotic cardiovascular disease
  - Severe (80%) focal narrowing of the left anterior descending coronary artery
  - Moderate (70%) focal narrowing of the left circumflex coronary artery
  - Moderate (50%) focal narrowing of the right coronary artery
  - No evidence of recent thrombus (blood clot in the coronary arteries)
- Hypertensive cardiomyopathy
  - Cardiomegaly (enlarged heart) (heart weighed 520 grams [g]; predicted normal weight between 296 g and 516 g as a function of sex, age, and body weight) [Silver and Silver 2001]
  - Left ventricular hypertrophy (LVH)
    Left ventricular wall and interventricular septum thickened (2.0 centimeters [cm] and 1. cm respectively;
    - normal by autopsy 0.76–0.88 cm [Colucci and Braunwald 1997];
    - normal by echocardiography 0.6–1.1 cm [Armstrong and Feigenbaum 2001]
    Microscopic evidence of myocyte hypertrophy, myocyte disarray, and interstitial fibros
- Normal cardiac valves
- No evidence of a pulmonary embolus (blood clot in the lung arteries)
- Carboxyhemoglobin (blood test for carbon monoxide exposure) 4.6% (normal for smokers; toxic level is 15% – 35% saturation [Winek 1976])
- Negative blood tests for drugs and alcohol

REFERENCES

Armstrong WF, Feigenbaum H [2001]. Echocardiography. In: Braunwald E, Zipes DP, Libby P, e Heart disease: a text of cardiovascular medicine. 6th ed. Vol. 1. Philadelphia, PA: W.B. Saunde Company, p. 167.

Colucci WS, Braunwald E [1997]. Pathophysiology of heart failure. In: Braunwald, ed. Heart disea 5th ed. Philadelphia, PA: W.B. Saunders Company, p. 401.

Silver MM, Silver MD [2001]. Examination of the heart and of cardiovascular specimens in surgical pathology. In: Silver MD, Gotlieb AI, Schoen FJ, eds. Cardiovascular pathology. 3rd e Philadelphia, PA: Churchill Livingstone, pp. 8–9.

Winek CL [1976]. Tabulation of therapeutic, toxic, and lethal concentrations of drugs and chemicals in blood. Clin Chem *22*(6):832–836.

Fatality Assessment and Control Evaluation
Investigation Report • F2009–29

Lieutenant Suffers Fatal Heart Attack During Fire Operations – Pennsylvania

The National Institute for Occupational Safety and Health (NIOSH), an institute within the Centers for Disease Control and Prevention (CDC), is the federal agency responsible for conducting research and making recommendations for the prevention of work-related injury and illness. In fiscal year 1998, the Congress appropriated funds to NIOSH to conduct a fire fighter initiative. NIOSH initiated the Fire Fighter Fatality Investigation and Prevention Program to examine deaths of fire fighters in the line of duty so that fire departments, fire fighters, fire service organizations, safety experts and researchers could learn from these incidents. The primary goal of these investigations is for NIOSH to make recommendations to prevent similar occurrences. These NIOSH investigations are intended to reduce or prevent future fire fighter deaths and are completely separate from the rulemaking, enforcement and inspection activities of any other federal or state agency. Under its program, NIOSH investigators interview persons with knowledge of the incident and review available records to develop a description of the conditions and circumstances leading to the deaths in order to provide a context for the agency's recommendations. The NIOSH summary of these conditions and circumstances in its reports is not intended as a legal statement of facts. This summary, as well as the conclusions and recommendations made by NIOSH, should not be used for the purpose of litigation or the adjudication of any claim. For further information, visit the program website at

www.cdc.gov/niosh/fire/
or call toll free
1–800–CDC–INFO (1–800–232–4636)

Death in the line of duty...

A summary of a NIOSH fire fighter fatality investigation June 11, 2009

Volunteer Fire Fighter Dies While Lost in Residential Structure Fire- Alabama

Incident Scene after Victim Removed
(Photo courtesy of sheriff's office)

SUMMARY

On October 29, 2008, a 24-year old male volunteer fire fighter (the victim) was fatally injured while fighting a residential structure fire. The victim, one of three fire fighters on scene, entered the

The National Institute for Occupational Safety and Health (NIOSH), an institute within the Centers for Disease Control and Prevention (CDC), is the federal agency responsible for conducting research and making recommendations for the prevention of work-related injury and illness. In fiscal year 1998, the Congress appropriated funds to NIOSH to conduct a fire fighter initiative. NIOSH initiated the Fire Fighter Fatality Investigation and Prevention Program to examine deaths of fire fighters in the line of duty so that fire departments, fire fighters, fire service organizations, safety experts and researchers could learn from these incidents. The primary goal of these investigations is for NIOSH to make recommendations to prevent similar occurrences. These NIOSH investigations are intended to reduce or prevent future fire fighter deaths and are completely separate from the rulemaking, enforcement and inspection activities of any other federal or state agency. Under its program, NIOSH investigators interview persons with knowledge of the incident and review available records to develop a description of the conditions and circumstances leading to the deaths in order to provide a context for the agency's recommendations. The NIOSH summary of these conditions and circumstances in its reports is not intended as a legal statement of facts. This summary, as well as the conclusions and recommendations made by NIOSH, should not be used for the purpose of litigation or the adjudication of any claim. For further information, visit the program website at www.cdc.gov/niosh/fire or call toll free at 1-800-CDC-INFO (1-800-232-4643).

Fatality Assessment and Control Evaluation Investigation Report # F2008-34

Volunteer Fire Fighter Dies While Lost in Residential Structure Fire- Alabama

residential structure by himself through a carport door with a partially charged 1½-in hose line; he became lost in thick black smoke. The victim radioed individuals on the fireground to get him out. Fire fighters were unable to locate the victim after he entered the structure which became engulfed in flames. The victim was caught in a flashover and was unable to escape the fire. Approximately an hour after the victim entered the structure alone, a police officer looking through the kitchen window noticed the victim's hand resting on a kitchen counter; the victim was nine feet from the carport door he had entered. The victim was removed from the structure and pronounced dead at the scene by emergency medical services. Key contributing factors identified in this investigation include: fire fighters entering a structure fire without adequate training, insufficient manpower, and lack of an established incident command system.

NIOSH investigators concluded that, to minimize the risk of similar occurrences, fire departments should:

- *ensure that fire fighters receive essential training consistent with national consensus standards on structural fire fighting before being allowed to operate at a fire incident*

- *develop, implement, and enforce written standard operating procedures (SOPs) for fireground operations*

- *ensure that fire fighters are trained to follow the two-in/two-out rule and maintain crew integrity at all times*

- *ensure that adequate numbers of apparatus and fire fighters are on scene before initiating an offensive fire attack in a structure fire*

- *ensure that officers and fire fighters know how to evaluate risk versus gain and perform a thorough scene size-up before initiating interior strategies and tactics*

- *develop, implement, and enforce a written incident management system to be followed at all emergency incident operations and ensure that officers and fire fighters are trained on how to implement the incident management system*

- *ensure fire fighters are trained in essential self-contained breathing apparatus (SCBA) and emergency survival skills*

- *ensure that protocols are developed on issuing a Mayday so that fire fighters and dispatch centers know how to respond*

- *ensure that a properly trained incident safety officer (ISO) is established at structure fires*

- *ensure that a rapid intervention team (RIT) is established and available at structure fires*

Fatality Assessment and Control Evaluation
Investigation Report # F2008-34

Volunteer Fire Fighter Dies While Lost in Residential Structure Fire- Alabama

- *ensure that properly coordinated ventilation is conducted on structure fires*

- *ensure that driver/pump operators receive adequate training to operate and maintain a water supply to hoselines on the fireground*

- *ensure that all fire fighters engaged in fireground activities wear the full array of personal protective equipment (PPE) issued to them*

- *ensure that fire fighters are trained to react to PASS and SCBA low air alarms, and that procedures are developed to properly shut down and secure a SCBA and its PASS device*

Additionally, states, municipalities, and authorities having jurisdiction

- *should consider requiring mandatory training for fire fighters*

INTRODUCTION

On October 29, 2008, a 24-year-old male volunteer fire fighter (the victim) died in a residential structure fire. On October 30, 2008, the U.S. Fire Administration notified the National Institute for Occupational Safety and Health (NIOSH) of this incident. On November 3-7, 2008, two safety and occupational health specialists from the NIOSH Fire Fighter Fatality Investigation and Prevention Program investigated this incident. The NIOSH investigators interviewed the officers and fire fighters of the volunteer departments involved in this incident and county EMS responders. The investigators also spoke with representatives from the Alabama State Fire College and the Alabama Association of Volunteer Fire Departments. The investigators met with the Deputy State Fire Marshal, sheriff's office investigator and the County 911 Dispatch Director. NIOSH investigators also reviewed witness statements and photographs of the fireground and dispatch tapes, the victim's training records, and the coroner's cause of death notification. The incident site was visited and photographed.

Although the performance of the victim's SCBA was not considered a factor in this incident, the SCBA was examined by NIOSH's National Personal Protective Technology Laboratory to determine conformity to the NIOSH approved configuration. When finalized, a summary of this evaluation will be added as an appendix to this report. At the request of the fire department, NIOSH contracted with a personal protective equipment (PPE) expert to evaluate the victim's PPE.[a] The expert evaluation concluded that the PPE was extensively damaged due to flame and heat exposures, most of which likely occurred after the victim succumbed to either smoke inhalation or severe burn injuries. It was

---

[a] The PPE evaluation report is available upon request to the NIOSH Division of Safety Research, Fire Fighter Fatality Investigation and Prevention Program (Attention: Tim Merinar), 1095 Willowdale Road, MS H1808, Morgantown, WV, 26505, 304-285-5916, Tmerinar@cdc.gov.

Fatality Assessment and Control Evaluation
Investigation Report # F2008-34

Volunteer Fire Fighter Dies While Lost in Residential Structure Fire- Alabama

not possible to identify if any of the PPE damage preceded the fatality, or if improper wearing of the PPE contributed to the ultimately fatal injuries. Where it was possible to identify the manufacturer, style of product, and manufacturing date, the gear appeared to be relatively new and compliant with the latest editions of relevant standards. The expert noted that no protective clothing or equipment would be expected to provide adequate protection in the circumstances of this event in which the victim was possibly exposed to a flashover event and subject to flame and high heat for nearly an hour.

FIRE DEPARTMENTS

- *Station "Alpha" – Victim's Department.* The victim's volunteer department has one station and is comprised of 21 fire fighters. The department serves a population of approximately 14,000 in a geographical area of 25 square miles.

- *Station "Bravo" – Mutual Aid Department.* The volunteer department has one station and is comprised of 20 fire fighters. The department serves a rural population in a geographical area of 28 square miles.

- *Station "Charlie" – Mutual Aid Department.* The volunteer department has one station and is comprised of 18 fire fighters. The department serves a population of approximately 8,000 in an area of about 47 square miles. *Note: Bravo Fire Department was dispatched before Charlie Fire Department, but did not respond until after the victim was located.*

The victim's department had no verbal or written standard operating procedures for their members to follow.

TRAINING and EXPERIENCE

The 24 year-old victim had been a volunteer fire fighter with this department for 2 years. The victim had attended documented peer-led training on self-contained breathing apparatus (SCBA), pump operations, water tactics, and general firefighting. The victim had also completed various online training courses on the incident command system (ICS) and national incident management system (NIMS).

The fire fighter initially operating with the victim had joined the fire department three months prior to the incident with no previous experience. Other fire fighters on scene had completed the same online courses in ICS and NIMS.

Alabama has no state training requirements for volunteer fire fighters. The Alabama State Fire College has a non-mandatory 160-hour volunteer fire fighter certification course.[1,2] Alpha fire

Fatality Assessment and Control Evaluation Investigation Report # F2008-34

Volunteer Fire Fighter Dies While Lost in Residential Structure Fire- Alabama

department members on scene had not completed this training. The three responding members from Charlie fire department had taken this training.

EQUIPMENT and PERSONNEL

- *Station "Alpha" - Victim's Department*
  Alpha Rescue 1 (AR1) with one fire fighter (fire fighter #1)
  Alpha Engine 2 (AE2) with two fire fighters (victim, fire fighter #2)
  Alpha Truck 2 (AT2) with one fire fighter (fire fighter #3)
  Alpha Engine 1 (AE1) with one fire fighter (fire fighter #4)

- *Station "Charlie" - Mutual Aid Department*
  Charlie Engine 3 (CE3) with fire chief (CFC) and one fire fighter (fire fighter #5)
  Charlie Engine 1 (CE1) with one fire fighter (fire fighter #6)
  Privately Owned Vehicle (POV) with one fire fighter

- *Water Supply on scene included*:
  AR1 300 gallons (used after last communication with victim)
  AE1 1,000 gallons (arrived after last communication with victim)
  AE2 1,000 gallons
  AT2 1,200 gallons (not used)
  CE1 1,000 gallons (arrived after last communication with victim)
  CE3 3,000 gallons (arrived after last communication with victim)

TIMELINE

The timeline for this incident includes the initial call to the 911 dispatch center at 1301 hours. Only the units directly involved in the operations preceding the incident are discussed in this report. Certain key radio transmissions are summarized in the timeline. All times are approximate. The response, listed in order of arrival, fire conditions and key events, includes:

- **1301 Hours**
  911 dispatch center receives a cellular 911 call for an attic fire in a house with all occupants out
  Alpha Fire Department dispatched

- **1310 Hours**
  AR1 en route

**FatalityAssessment and Control Evaluation
Investigation Report # F2008-34**

Volunteer Fire Fighter Dies While Lost in Residential Structure Fire- Alabama

- **1316 Hours (thick black smoke from roof)**
  AR1 on scene and states, "smoke coming from roof"
  AR1 driver requests Bravo Fire Department to be dispatched
  Bravo Fire Department dispatched

- **1317 Hours**
  AE2, operated by the victim, on scene (no en route time available)
  County EMS dispatched

- **1318 Hours**
  Bravo Fire Department dispatched again after no response
  County EMS en route
  Victim possibly requesting his hoseline to be charged

- **1319 Hours**
  AE1 en route
  FF1 advises victim he is charging his line

- **1320 Hours**
  Sounds like victim states, "...fire getting away in here..."

- **1323 Hours**
  Charlie Fire Department dispatched after Bravo Fire Department was dispatched twice with no response
  Inaudible radio traffic by victim

- **1324 Hours**
  Sounds like victim yells "I'm hot...Come get me, come get me!"

- **1328 Hours (fire blowing out "A" side windows, front door, and carport entry)**
  AT2 on scene (no en route time available)
  AE1 on scene

- **1330 Hours**
  CE1 and CE3 en route
  County EMS on scene

- **1333 Hours**
  AR1 requested ambulance
  911 dispatch center advised AR1 that ambulance may be on scene
  AR1 requests second ambulance to be dispatched
  911 dispatch center requested reason for a second ambulance, with no response

Fatality Assessment and Control Evaluation Investigation Report # F2008-34

Volunteer Fire Fighter Dies While Lost in Residential Structure Fire- Alabama

- **1338 Hours (fire visible from doors and "A" and "B" side windows)**
  CE3 on scene

- **1339 Hours**
  CE1 on scene

- **1356 Hours**
  AR1 requested local power company to be contacted

- **1418 Hours**
  911 dispatch center received cellular 911 call from police officer on scene requesting Bravo and Delta Fire Departments be dispatched to the scene to help with a trapped fire fighter

- **1420 Hours**
  Victim found and removed from structure

PERSONAL PROTECTIVE EQUIPMENT

The victim was last seen wearing a full array of personal protective clothing and equipment, consisting of turnout gear (coat and pants), helmet, Nomex® hood, gloves, boots, and a self-contained breathing apparatus (SCBA) with an integrated personal alert safety system (PASS). The structural fire fighting gear was compliant with the 2007 edition of NFPA 1971. The victim was equipped with a portable radio, flashlight, and various fire fighter hand tools in his pockets. The heat resistant outer shell, moisture barrier, and insulating thermal lining were all present during the incident and documented during the investigation. The victim was found without his helmet and Nomex® hood on. The victim was also missing a glove, boot, and assigned radio. The face piece appeared to have been melted off of the victim.

STRUCTURE

The incident structure was a single-story brick ranch house built in 1969. The residential structure had approximately 2,100 square feet of furnished living area and no basement. The interior construction consisted of wood framing and possibly drywall. The exterior construction was brick with an attached carport at the "A-D" corner (see Diagram). A tin roof had been placed overtop the existing shingled roof.

The origin and cause of the fire was ruled accidental by the Deputy State Fire Marshal and believed to have started in the chimney. The fireplace had been converted into a wood stove with a flue in or around 1970 or 1971. Approximately 8-10 years prior to the incident, the brick chimney had begun to pull away from the house. The owner had wrapped a steel cable around the chimney and placed a

Fatality Assessment and Control Evaluation
Investigation Report # F2008-34

Volunteer Fire Fighter Dies While Lost in Residential Structure Fire- Alabama

turnbuckle to connect the cable ends; this supported the chimney while pulling it back against the house (see Photo 1).

Photo 1. Steel cable used to support the chimney against the house with a turnbuckle.
(Photos courtesy of sheriff's office)

After the chimney was repositioned, the owner noticed that bricks from the chimney's flue had cracked and shifted. The owner removed the damaged bricks and replaced them with hollow cinder blocks. These blocks had been laid with the hollows horizontally positioned towards the attic space (see Photo 2). The residents of the house were safely evacuated prior to the fire department's arrival. The structure was completely destroyed by a rekindle several days later.

Photo 2. The picture shows the area of the brick chimney flue that was replaced with cinder blocks. The cinder blocks allowed heat and flame to impinge on the exterior wall allowing fire to spread into the attic space.
(Photo courtesy of sheriff's office)

Fatality Assessment and Control Evaluation Investigation Report # F2008-34

Volunteer Fire Fighter Dies While Lost in Residential Structure Fire- Alabama

WEATHER

The weather at the time of the incident was clear with a temperature of 55°F and slight winds from the west.

INVESTIGATION

On October 29, 2008, at 1301 hours, the 911 dispatch center received a cellular 911 call for an attic fire. The initial dispatch included Station Alpha at 1301 hours, Station Bravo at 1316 hours (no response), and Station Charlie at 1323 hours. Smoke was showing from the roof upon arrival of Station Alpha's first unit. Incident command was not established by arriving units.

Initial activities of AR1 and AE2

AR1 marked on scene at 1316 hours with thick, black smoke showing from the roof. Fire fighter #1 (FF1) exited the apparatus after he had positioned it in the street in front of the house. He then spoke to several individuals in the yard that had appeared to be exiting away from the house. A female advised him that there was no one left in the house. At 1317 hours the victim and fire fighter #2 (FF2) marked on scene in AE2 and positioned their apparatus behind AR1 (see Diagram). FF1 met the victim and FF2 between their apparatus to put gear on. The victim and FF2 then walked around the house to check on fire conditions. FF1 then flaked out cross-lay #1 (200-ft of 1½-in hose) from AE2. FF2 returned to AE2 to retrieve the hoseline while the victim waited at the A-D corner under the carport. It is believed that the victim and FF2 were on air. FF2 pulled the uncharged cross-lay #1 to the carport door and handed the nozzle to the victim, while FF1 charged the line. The victim opened the bail and water trickled out of the nozzle. The victim yelled back to FF1 to give him more pressure on the hoseline. FF2 recalls hearing the engine of AE2 get really loud, but could not confirm the line being fully charged. The victim and FF2, on air, walked into the structure through the carport door. They were approximately two feet inside the structure and were met by thick, rolling black smoke, but no fire. Quickly, they exited through the carport door taking cross-lay #1 with them. The victim told FF2 to go and get a flashlight. FF2 reported that he told the victim to wait for his return, and that the victim indicated that he would wait.

Activities of FF2

FF2 reported that he walked to the end of the driveway (approximately 75-ft) grabbed the flashlight off of the apparatus, and returned to the carport door. The victim was no longer under the carport, thick black smoke was rolling from the door, and cross-lay #1 was stretched into the structure (see Photo 3). FF2, still on air, entered back into the house through the carport door but could not see his hands or feet just inside the door. He turned the flashlight on and still could not see anything but rolling black smoke. He started yelling the victim's name and listening for the sound of water flowing or a PASS device. He heard nothing before his low air alarm sounded. *Note: FF1 stated his low air alarm sounded after 15-20 minutes of being on air.*

NIOSH Fire Fighter Fatality Investigation and Prevention Program — Fatality Assessment and Control Evaluation Investigation Report # F2008-34

Volunteer Fire Fighter Dies While Lost in Residential Structure Fire- Alabama

Photo 3. Door under carport that victim entered through.
(Photo courtesy of sheriff's office)

Activities of FF1

Moments after FF2 left with the flashlight, FF1 reported hearing the victim radio him and stating, "…at front door, can't get it opened." *Note: This was not heard on the 911 dispatch tape by NIOSH investigators and is believed to have been on a non-recorded fire department talk around channel.* FF1 retrieved an axe from AE2 and donned his SCBA over his bunker gear. He ran to the front porch and found the screen and exterior doors locked. He yanked the screen door off the frame and punched the exterior door open with the head of the axe. He was met with thick black smoke causing him to step back and clean the soot from his face piece. He then stood in the doorway yelling for the victim. FF1 then heard glass breaking from the bay window beside the front door. This window was between the front door and carport (see Photo 4).

Photo 4. Incident scene after rekindle and collapse showing the front door and bay window.
(NIOSH photo)

Volunteer Fire Fighter Dies While Lost in Residential Structure Fire- Alabama

FF1 thought it was the victim trying to get out. He used the axe and removed the rest of the window pane from the frame. He yelled for the victim again from this window with no response. He then walked off the porch to the carport to see if the victim had come out of the house. He then ran down the D-side to the C-side of the house to look for a back door. During this exterior search, the victim radioed FF1 again, "get me out!" With no available access from C-side, FF1 ran back to the A-side of the house and was met by FF2.

Activities of FF1 and FF2
They both spoke briefly about the victim being lost inside the house. FF2 asked FF1 to assist him in changing his air bottle. After getting his bottle changed, FF2 attempted to get cross-lay #2 (200-ft of 1½-in hose) off AE2 when he noticed the house was "engulfed" in flames. *Note: Fire was believed to be pushing out the carport door and A-side bay window.* FF1 radioed the 911 dispatch center and requested a second ambulance. *Note: The first ambulance was dispatched to check on the occupants of the house and was available on scene.* The dispatcher asked him why he needed a second ambulance, but never got a response. FF1 noticed that cross-lay #1 was flat so he shut the line down, disconnected the section of hose closest to the carport not exposed to flames, and retrieved a nozzle from the arriving AT2 with fire fighter #3 (FF3). *Note: This was the same line that the victim had taken into the house. During the fire investigation it was discovered that the hoseline was burnt through at the carport threshold.* FF1 then assisted FF2 in flaking out cross-lay #2 to the house. FF1 saw that AE1 had arrived with fire fighter #4 (FF4). He told FF4 about the victim missing in the structure and the inability to get him out. After cross-lay #1 was disconnected from the burnt section of hose and a new nozzle connected, it was placed back in service. FF1 took cross-lay #1 and sprayed water on the front porch until AE2 ran out of water. When AE2 ran out of water, FF1 dropped cross-lay #1 in the front yard and drove AE1 a ½ mile down the road to a hydrant to fill up. Arriving apparatus from Charlie Fire Department then supplied AE2.

Activities of AT2 and AE1
AT2 and AE1 marked on scene at 1328 hours with fire blowing out the windows on the A-side, front door and carport entry. FF4 set AE1 to pump and supplied their 1,000 gallons of water to AE2 via a 2½-in supply line. AT2 did not supply or receive any water to or from apparatus on scene. No hoselines were stretched from AT2 or AE1. FF4 monitored the pump panels of AE1 and AE2 while FF3 briefly assisted FF2 with charging cross-lay #2. FF3 then picked up cross-lay #1 and took it with him as he entered the structure briefly via the front door. He briefly sprayed water as he yelled for the victim; he heard neither a response nor a sounding PASS device. He was quickly pushed back through the front door by intense heat and fire. FF3 made a second entry through the front door to locate the victim with no success.

Activities of CE1 and CE3
CE3 marked on scene at 1338 hours with their fire chief (CFC) and fire fighter #5 (FF5); the CFC stated the house was "heavily involved" with fire. The fire had not vented through the roof, but was visible from A-side and B-side window and door openings. AE1 advised the CFC that he was going to get water. FF5 took a 2½-in supply line from CE3 and supplied AE2 with its' 3,000 gallons of

Fatality Assessment and Control Evaluation
Investigation Report # F2008-34

Volunteer Fire Fighter Dies While Lost in Residential Structure Fire- Alabama

water. FF5 operated the pump panel on CE3 with assistance from the CFC. The CFC pulled one preconnected 150-ft 1¾-in hoseline to the front yard and one 100-ft 1¾-in hoseline to the A-D corner. CE1 marked on scene at 1339 hours with fire fighter #6 (FF6). FF6 exited his apparatus and briefly spoke to FF4 about what was going on. FF6 then picked up what is believed to have been cross-lay #2 that was placed on the ground and not in use. He stated the line had too much pressure on it for him to handle and that he did not see anyone operating the pump panel on AE2 to fix it. *Note: When interviewed, one pump operator stated the department did not have a set pump discharge pressure for different or multiple lines run from an apparatus. A pump operator would adjust the pressure according to fire fighters operating the hand line.* He also noticed a 1-in booster line from AR1 in use by Alpha Fire Department members on the A-side. The CFC took the 2½-in supply line from CE1 and hooked it into CE3. The CFC then operated a hoseline at the A-D corner while FF6 operated a hand line at the A-B corner (not clear which apparatus the line was pulled from). A third fire fighter from Charlie Fire Department arrived on scene in his (POV) and assisted a fire fighter with flowing water into the A-side bay window. The fire fighters on CE1 and CE3 were unable to make an interior attack or get close to the burning house because they had responded to the incident without their structural fire fighting gear. *Note: All Charlie Fire Department members on scene of this incident responded without structural firefighting gear. Their gear had been taken to their residences in case they responded to incidents from home, instead of responding to the fire station to pick up an apparatus. This was a common practice by the Fire Chief and his members.* When CE1 and CE3 tank water ran out, FF6 and CFC left in CE1 and CE3 to find the hydrant ½ mile down the road. When they returned, the victim was being removed from the structure. By this time, AE1 had returned and was positioned at the B-C corner and AE1 was again, out of water.

Victim Discovery
Individual responders interviewed by NIOSH remarked on several air packs taken off by fire fighters and left unattended on the fireground. These air packs were left on and the activated integrated PASS devices continued to alarm. This may have hindered the efforts of fire fighters to locate the victim during rescue efforts because they could not determine whether the PASS alarm was coming from inside or outside the structure. A town police officer who had been on scene since the initial dispatch was assisting fire fighters in trying to locate the victim from the exterior. FF4 came upon a burnt glove on the ground in front of the C-side kitchen window. The police officer grabbed an extension ladder and placed it against the window sill of the kitchen window. The police officer stated the fire had burned itself out in this area and the smoke was light enough to get a good view inside the house. He noticed the victim's hand resting on the kitchen counter under some debris. He immediately alerted fire fighters who were close by. Two fire fighters entered through the carport door, into the laundry room and turned left into the kitchen where they saw the police officer in the window. The victim was found on the kitchen floor in front of the kitchen sink; he was removed from the house and pronounced dead on the scene by County EMS responders at 1420 hours. A nozzle connected to the original cross-lay #1, the victim's radio, helmet, face piece, SCBA, and a boot were found scattered around him (see Photo 5). It appeared that the helmet, face piece, and SCBA were burned off of the victim. The nozzle was found with the bale fully opened at a 30 degree fog setting.

Fatality Assessment and Control Evaluation
Investigation Report # F2008-34

Volunteer Fire Fighter Dies While Lost in Residential Structure Fire- Alabama

1. Victim – 112" from carport door
2. SCBA – 85" from carport door
3. Face Piece – 128" from carport door
4. Helmet – 139 ½" from carport door
5. Radio – 147" from carport door
6. Boot – 94" from carport door
7. Nozzle – 101" from carport door

Photo 5. Number cues placed to document victim and items discovered.
(Photo courtesy of sheriff's office)

Fire Behavior and Spread

During their investigation, the fire marshal and sheriff's office investigators believed that the chimney flue failed allowing radiant heat and flame to travel through the open sides of the cinder blocks and impinge on combustible materials in the attic. It is believed that the fire smoldered for a while in the attic before finding a reliable fuel source. Fire burned across the C-side of the structure's attic before burning down into the living area where the wood stove was located (see Photo 6). Thick black smoke was filling the structure when the victim and FF2 entered through the carport for the first time. No flames were immediately visible according to FF2. FF1 was met with the same smoke conditions when he opened the front door and the bay window broke under extreme heat conditions. Flames throughout the house erupted moments later. FF1 and FF2 do not recall any type of explosion or "puffs" of smoke before the house erupted in flames. The sheriff's office investigator believed that the house "flashed," trapping the victim and exposing him to extreme heat and fire conditions after the door and window were opened, which would have allowed fresh air (oxygen) to join the fuel mixture (heavy dark smoke) inside, ultimately leading to the flashover.

Fatality Assessment and Control Evaluation Investigation Report # F2008-34

Volunteer Fire Fighter Dies While Lost in Residential Structure Fire- Alabama

Photo 6. Fire damage to living room area before rekindle and structural collapse.
(Photo courtesy of sheriff's office)

CONTRIBUTING FACTORS

Occupational injuries and fatalities are often the result of one or more contributing factors or key events in a larger sequence of events that ultimately result in the injury or fatality. NIOSH investigators identified the following items as key contributing factors in this incident that ultimately led to the fatality:

- Fire fighters operating on a fireground and entering a burning structure without adequate training.
- Insufficient manpower to combat the fire.
- No incident command system established.

Fatality Assessment and Control Evaluation Investigation Report # F2008-34

Volunteer Fire Fighter Dies While Lost in Residential Structure Fire- Alabama

CAUSE OF DEATH

According to the county medical examiner's office, the victim died from smoke inhalation and thermal burns. The victim's carboxyhemoglobin (COHb) was 35%.

RECOMMENDATIONS

Recommendation #1: Fire departments should ensure that fire fighters receive essential training consistent with national consensus standards on structural fire fighting before being allowed to operate at a fire incident.

Discussion: Training on structural fire fighting is essential for fire fighter safety and survival. This training should include, but not be limited to, departmental standard operating procedures, fire fighter safety, building construction, fire behavior, and fireground tactics. NFPA 1001 *Standard for Fire Fighter Professional Qualifications* was established to facilitate the development of nationally applicable performance standards for fire service personnel.[3] NFPA 1500 *Fire Department Occupational Safety and Health Program*, Chapter 5, requires that the fire department provide an annual skills check to verify minimum professional qualifications of its members.[4] The purpose of NFPA 1001 *Standard for Fire Fighter Professional Qualifications* is to show clear and concise requirements that can be used to determine that an individual, when measured to the standard, possesses the knowledge, skills, and abilities to perform as a fire fighter and that these requirements can be used by any fire department in the country. Once the basic skills and knowledge sets of Fire Fighter I are met the individual can continue on to Fire Fighter II. Staying proficient at these levels is only possible through a training regime developed by the fire department or training entity that covers topics like ventilation, hazard recognition, fire behavior, incident command system, scene size-up, and basic water operations.

The state of Alabama does offer a non-mandatory 160 hour volunteer fire fighter certification course. This course is optional and not required to be a volunteer fire fighter in the state of Alabama. Neither the victim nor fire fighters from his department on scene had completed this training. During interviews, the fire chief advised NIOSH investigators that the victim and two other fire fighters were scheduled to start this training in January 2009. The chief currently has nine fire fighters in the 160 hour certification program that are scheduled to graduate in June 2009. This basic training provides fire fighters with the knowledge, skills, and abilities to make sound, safe decisions before engaging in active fire suppression. Fire departments should pair untrained and inexperienced fire fighters with a trained and experienced fire fighter.

Recommendation #2: Fire departments should develop, implement, and enforce written standard operating procedures (SOPs) for fireground operations.

Discussion: Written SOPs enable individual fire department members an opportunity to read and

Fatality Assessment and Control Evaluation Investigation Report # F2008-34

Volunteer Fire Fighter Dies While Lost in Residential Structure Fire- Alabama

maintain a level of assumed understanding of operational procedures. Conversely, fire departments can suffer when there is an absence of well developed SOPs. The NIOSH Alert, *Preventing Injuries and Deaths of Fire Fighters* identifies the need to establish and follow fire fighting policies and procedures.[5] Guidelines and procedures should be developed, fully implemented and enforced to be effective. The following NFPA Standards also identify the need for written documentation to guide fire fighting operations:

NFPA 1500 *Fire Department Occupational Safety and Health Program* states that fire departments shall prepare and maintain policies and standard operating procedures that document the organizational structure, membership, roles and responsibilities, expected functions, and training requirements, including the following….(4) The procedures that will be employed to initiate and manage operations at the scene of an emergency incident.[4]

NFPA 1561 *Standard on Emergency Services Incident Management System* states that standard operating procedures (SOPs) shall include the requirements for implementation of the incident management system and shall describe the options that are available for application according to the needs of each particular situation.[6]

NFPA 1720 *Standard for the Organization and Deployment of Fire Suppression Operations, Emergency Medical Operations, and Special Operations to the Public by Volunteer Fire Departments* states that the authority having jurisdiction shall promulgate the fire department's organizational, operational, and deployment procedures by issuing written administrative regulations, standard operating procedures, and departmental orders.[7]

The victim's fire department had not implemented any verbal or written SOPs for their members. An effective SOP can aid in the decision making process when on the fireground.

Recommendation #3: Fire departments should ensure that fire fighters are trained to follow the two-in/two-out rule and maintain crew integrity at all times.

Discussion: NFPA 1500 *Fire Department Occupational Safety and Health Program* states that the two-in/two-out rule should be used when making entry into a hazardous area.[4] The team members should be in communication with each other through visual, audible, or electronic means to coordinate all activities, and determine if emergency rescue is needed. Working alone in a structure fire does not provide a fire fighter a back-up plan if he or she gets in trouble.

Crew integrity relies on knowing your team members and the team leader, maintaining visual contact (if visibility is low, teams must stay within touch or voice distance of one another), communicating needs and observations to the team leader, rotating to rehab, staging as a team, and watching fellow team members by means of a buddy system. Crew integrity is being able to maintain a cohesive crew over a period of time. Fire fighter accountability is an important aspect of fire ground safety that can be compromised when teams are split up. Being able to operate as a crew and understanding one's

Fatality Assessment and Control Evaluation Investigation Report # F2008-34

Volunteer Fire Fighter Dies While Lost in Residential Structure Fire- Alabama

limitations will benefit the crew's safety and overall incident outcome. This is especially important during interior fire attack. Names on coats, reflective shields or company numbers on helmets, and helmet and turnout clothing colors are visual ques that fire fighters can use to maintain crew integrity in poor visibility.

During this incident, the victim entered the house alone and was unable to self-rescue or be rescued by on scene fire fighters.

Recommendation #4: Fire departments should ensure that adequate numbers of apparatus and fire fighters are on scene before initiating an offensive fire attack in a structure fire.

Discussion: Fire suppression operations should be organized to ensure that the fire department's fire suppression capability includes sufficient personnel, equipment, and other resources to deploy fire suppression resources efficiently, effectively, and safely.[7] Volunteer fire departments need to identify minimum staffing requirements to ensure that an adequate number of members are available to operate safely and effectively. Rural areas have a lower population density and require at least six people (two-in/two-out plus the incident commander and pump operator) on scene before fire suppression operations can take place.[7] If staffing is known to be limited during the day, then automatic mutual aid agreements should be implemented with surrounding departments to ensure a timely response of resources. Fire departments should develop and establish good working relationships with surrounding departments so that reciprocal assistance and mutual aid is readily available when emergency situations escalate beyond response capabilities. NFPA 1720 *Standard for the Organization and Deployment of Fire Suppression Operations, Emergency Medical Operations, and Special Operations to the Public by Volunteer Fire Departments* states that fire departments should have the capability for sustained operations to include:

- fire suppression
- engagement in search and rescue
- forcible entry
- ventilation
- preservation of property
- accountability for personnel
- dedicated rapid intervention team (RIT)
- provision for support activities for situations that are beyond the capability of the initial attack

During this incident, the initial responding units consisted of a rescue truck with one fire fighter and an engine with the victim and a fire fighter. The next due mutual aid station did not have any available personnel to respond to the incident scene. This level of initial response did not meet what is outlined in NFPA 1720 *Standard for the Organization and Deployment of Fire Suppression Operations, Emergency Medical Operations, and Special Operations to the Public by Volunteer Fire Departments* and is insufficient for fire fighting operations. The fire department had not outlined

FatalityAssessment and Control Evaluation
Investigation Report # F2008-34

Volunteer Fire Fighter Dies While Lost in Residential Structure Fire- Alabama

minimum staffing procedures for departmental responses. In rural areas and areas with long response times, automatic mutual aid should be established to ensure enough fire fighters arrive in a timely manner to safely perform fireground tasks. Having only three fire fighters to sustain operations on the fireground negatively affected the control of the fire and overwhelmed the fire fighters that were on scene.

Recommendation #5: Fire departments should ensure that officers and fire fighters know how to evaluate risk versus gain and perform a thorough scene size-up before initiating interior strategies and tactics.

Discussion: Situational awareness is a highly critical aspect of human decision making: the understanding of what is happening around you, projecting future situation events, comprehending information and its relevance, and an individual's perception.[8] This is especially important when conducting an initial size-up. The initial size-up of the incident helps to determine the number of fire fighters and the amount of apparatus and equipment needed to control the fire, assists in determining the most effective point of fire attack, venting heat and smoke, and whether the attack should be offensive or defensive. The size-up should include an evaluation of factors such as the fire size and location, length of time the fire has been burning, conditions on arrival, occupancy, fuel load and presence of combustible or hazardous materials, exposures, time of day, and weather conditions.[7] Information on the structure itself including size, construction type, age, condition, evidence of renovations, lightweight construction, and loads on roof and walls will aid in determining strategies and tactics. Fire fighters need to consider warning signs like dense black smoke, turbulent smoke, smoke puffing around doorframes, discolored glass, and a reverse flow of smoke back inside the building before making entry into a structure fire.[9] The level of risk to the fire fighters must be balanced against the potential to save lives or property.[6]

During this incident, it was reported that all occupants from the residence were accounted for. There was no life threatening emergency on scene that would have required fire fighters to rush into the house before thinking about their risks and potential gains from combating a fire with limited staffing. Incident command was not established, fire fighters on scene had limited fire fighting training, manpower was insufficient for the incident, and all occupants of the house were accounted for prior to their arrival. A decision to combat this fire defensively until additional resources arrived was warranted. A flashover occurred trapping the victim within the structure and overwhelming the fire fighters and available resources on scene.

Recommendation #6: Fire departments should develop, implement, and enforce a written incident management system to be followed at all emergency incident operations and ensure that officers and fire fighters are trained on how to implement the incident management system.

Discussion: NFPA 1500 *Standard on Fire Department Occupational Safety and Health Program*[4] and NFPA 1561 *Standard on Emergency Services Incident Management System*[6] both state that an incident management system (IMS) should be utilized at all emergency incidents (including but not

Fatality Assessment and Control Evaluation
Investigation Report # F2008-34

Volunteer Fire Fighter Dies While Lost in Residential Structure Fire- Alabama

limited to training exercises). NFPA 1561, Chapter 3.3.29 defines the incident management system (also known as the incident command system (ICS)) as "a system that defines the roles and responsibilities to be assumed by responders and the standard operating procedures to be used in the management and direction of emergency incidents and other functions.[6] Chapter 4.1 states "the incident management system shall provide structure and coordination to the management of emergency incident operations to provide for the safety and health of emergency services organization (ESO) responders and other persons involved in those activities." Chapter 4.2 states "The incident management system shall integrate risk management into the regular functions of incident command." Each fire department or emergency services organization (ESO) should adopt an incident management system to manage all emergency incidents. The IMS should be defined and in writing and include standard operating procedure (SOPs) covering the implementation of the IMS. The IMS should include written plans that address the requirements of different types of incidents that can be anticipated in each fire department's or ESO's jurisdiction. The IMS should address both routine and unusual incidents of differing types, sizes and complexities. The IMS covers more than just fireground operations. The IMS must cover incident command, accountability, risk management, communications, rapid intervention crews (RIC), roles and responsibilities of the incident safety officer (ISO), and interoperability with multiple agencies (police, emergency medical services, state and federal government, etc.) and surrounding jurisdictions (mutual aid responders).

NFPA 1720 *Standard for the Organization and Deployment of Fire Suppression Operations, Emergency Medical Operations, and Special Operations to the Public by Volunteer Fire Departments* states, "the incident commander shall be responsible for the overall coordination and direction of all activities for the duration of the incident."[7] Furthermore, the incident command system/incident management system is a standardized on-scene incident management concept designed specifically to allow responders to adopt an integrated organizational structure equal to the complexity and demands of any single incident or multiple incidents without being hindered by jurisdictional boundaries.[10] The IC has several responsibilities upon his arrival including safety and health of responders on scene, stabilizing the incident, developing strategies, and overall management of the incident. Without an IC, the safety of the fire fighter and fireground operations can be compromised.

During this incident, no IC was established. The victim, FF1, and FF2 performed tasks that included a walk around of the house, stretching an attack line to an entry point, and setting up a water supply. Not having an IC to provide direction on managing the incident through sound strategies and tactics contributed to the fire getting out of control and fatally injuring the victim.

Recommendation #7: Fire departments should ensure that fire fighters are trained in essential self-contained breathing apparatus (SCBA) and emergency survival skills.

Discussion: Fire fighters must act promptly when they become lost, disoriented, injured, low on air, or trapped.[11-16] First, they must transmit a distress signal while they still have the capability and sufficient air, noting their location if possible. The next step is to manually activate their PASS

Fatality Assessment and Control Evaluation Investigation Report # F2008-34

Volunteer Fire Fighter Dies While Lost in Residential Structure Fire- Alabama

device. To conserve air while waiting to be rescued, fire fighters should try to stay calm, be focused on their situation and avoid unnecessary physical activity. They should survey their surroundings to get their bearings and determine potential escape routes such as windows, doors, hallways, changes in flooring surfaces, etc.; and stay in radio contact with the IC and other rescuers. Additionally, fire fighters can attract attention by maximizing the sound of their PASS device (e.g. by pointing it in an open direction), pointing their flashlight toward the ceiling or moving it around, and using a tool to make tapping noises on the floor or wall. A crew member who initiates a Mayday call for another person should quickly try to communicate with the missing member via radio and, if unsuccessful, initiate another Mayday providing relevant information on the missing fire fighter's last known location.

During this incident, it was never determined whether the victim's PASS device was heard. The victim radioed for assistance but did not declare a Mayday. The victim was found within arms reach of his hoseline that led outside (approximately eight feet to the door).

Recommendation #8: Fire departments should ensure that protocols are developed to assist fire fighters in issuing a Mayday so that fire fighters and dispatch centers know how to respond.

Discussion: A radio transmission reporting a trapped or downed fire fighter is the highest priority transmission that an IC can receive. Mayday transmissions must always be acknowledged and immediate action must be taken.[11,12] As soon as fire fighters become lost or disoriented, trapped or unsuccessful at finding their way out of the interior of a structural fire, they must initiate emergency radio transmissions. A Mayday call should receive the highest communications priority from dispatch, the IC, and all other units on-scene. Dispatchers should be trained to monitor radio traffic so that they can assist the IC in acknowledging distress calls. Dispatchers are not exposed to fireground noise like the IC, which may distract the IC from acknowledging radio transmissions.

During this incident, the victim made at least two calls for help inside the house. The department did not have procedures for ensuring that all on-scene fire fighters and the dispatch center were aware of the trapped fire fighter, that information crucial for locating the fire fighter was collected, and that organized rescue attempts were initiated. Individual attempts to find and rescue the victim were unsuccessful and 58 minutes passed from the victim's initial call for help and dispatch being notified of the victim being in trouble.

Recommendation #9: Fire departments should ensure that a properly trained incident safety officer (ISO) is established at structure fires.

Discussion: NFPA 1521 *Standard for Fire Department Safety Officer* defines the role of the ISO at an incident scene and identifies duties such as reporting pertinent fireground information to the IC; ensuring the department's accountability system is in place and operational; monitoring radio transmissions and identifying barriers to effective communications; and ensuring that established

Fatality Assessment and Control Evaluation Investigation Report # F2008-34

Volunteer Fire Fighter Dies While Lost in Residential Structure Fire- Alabama

safety zones, collapse zones and other designated hazard areas are communicated to all members on scene.[17] Although the presence of a safety officer does not diminish the responsibility of individual fire fighters and fire officers for safety, the ISO adds a higher level of attention and expertise to help the individuals. The ISO must have particular expertise in analyzing safety hazards and must know the particular uses and limitations of protective equipment.[6]

Having an ISO standard operating procedure and an available and trained individual as an ISO on this incident may have prevented the victim from going into the structure alone; and could have provided assistance in handling the victim's call for help.

Recommendation #10: Fire departments should ensure that a rapid intervention team (RIT) is established and available at structure fires.

Discussion: A RIT should be designated and available to respond before interior attack operations begin. The team should report to the incident commander and be available within the incident's staging area. The RIT should have all tools necessary to complete the job, e.g., search and rescue ropes, Halligan bar and flat-head axe combo, first-aid kit, and resuscitation equipment.[9] These teams can intervene quickly to rescue a fire fighter who is running out of breathing air, becomes disoriented, lost in smoke filled environments, trapped by fire, or involved in structural collapse.[4]

During this incident, there were only two other fire fighters on scene when the victim radioed for help. Neither one of them had received training on rapid intervention or basic fire fighting. NFPA 1500 *Standard on Fire Department Occupational Safety and Health Program*, Chapter 8.8, Rapid Intervention for Rescue of Members, provides detailed guidelines for the deployment of rescue teams at emergency incidents.[4] Chapter 8.8.1 states "The fire department shall provide personnel for the rescue of members operating at emergency incidents." The staffing requirements set forth by NFPA 1720 *Standard for the Organization and Development of Fire Suppression Operations, Emergency Medical Operations, and Special Operations to the Public by Volunteer Fire Departments* were not met; hence, not allowing them to establish a RIT or follow a two-in/two-out rule. *Note: Although volunteer fire departments in the State of Alabama are not governed by federal regulations set forth by the Occupational Safety and Health Administration, it is good practice that OSHA's two-in/two-out rule, 29 CFR 1910.134 (g)(4)(i) be adopted and written into fire department standard operating procedures.*

Recommendation #11: Fire departments should ensure that properly coordinated ventilation is conducted on structure fires.

Discussion: Ventilation is the systematic removal of heated air, smoke, and fire gases from a burning building and replacing them with cooler air.[9] Properly coordinated ventilation can decrease how fast the fire spreads, increase visibility, and lower the potential for flashover or backdraft. Proper ventilation reduces the threat of flashover by removing heat before combustibles in a room or enclosed area reach their ignition temperatures, and can reduce the risk of a backdraft by reducing the

Fatality Assessment and Control Evaluation Investigation Report # F2008-34

Volunteer Fire Fighter Dies While Lost in Residential Structure Fire- Alabama

potential for superheated fire gases and smoke to accumulate in an enclosed area. Properly ventilating a structure fire will reduce the tendency for rising heat, smoke, and fire gases, trapped by the roof or ceiling, to accumulate, bank down, and spread laterally to other areas within the structure.

The ventilation opening may produce a chimney effect causing air movement from within a structure toward the opening. This air movement helps facilitate the venting of smoke, hot gases and products of combustion, but may also cause the fire to grow in intensity and may endanger fire fighters who are between the fire and the ventilation opening. For this reason, ventilation should be closely coordinated with hoseline placement and offensive fire suppression tactics. Close coordination means the hoseline is in place and ready to operate so that when ventilation occurs, the hoseline can overcome the increase in combustion likely to occur. If a ventilation opening is made directly above a fire, fire spread may be reduced, allowing fire fighters the opportunity to extinguish the fire. If the opening is made elsewhere, the chimney effect may actually contribute to the spread of the fire.[9] The IC needs to consider the following and how it will affect ventilation and overall control of the fire:

- Who will ventilate (knowledge and skills)
- What type of ventilation
- When to ventilate
- Where to ventilate
- Why ventilate
- How to properly and safely ventilate

The two types of ventilation are vertical and horizontal. During vertical ventilation the natural convection of the heated gases creates upward currents that draw the fire and heat in the direction of the vertical opening. Horizontal ventilation allows for heat, smoke, and gases to escape by means of a doorway or window, but is highly influenced by the location and extent of the fire and should be cautioned if the fire is in the attic.[9]

During this incident, no attempt was made to ventilate the attic fire in the house. The original roof layers had been covered with a tin roof which would have hindered fire fighter's attempts to ventilate the roof. No windows were broken out on the first floor.

Recommendation #12: Fire departments should ensure that driver/pump operators receive adequate training to operate and maintain a water supply to hoselines on the fireground.

Discussion: NFPA 1002 *Standard for Fire Apparatus Driver/Operator Professional Qualifications* sets minimum qualifications for driver/operators. The pump operator is responsible for the life safety of all personnel exposed to dangerous situations that are operating hoselines being supplied by the pumper.[18] The qualities and skill sets needed by a pump operator include an understanding of different types of pumping apparatus, proper apparatus placement to maximize water supply efficiency, fire pump theory and operation, hydraulic calculations, and water supply choices. In

Fatality Assessment and Control Evaluation Investigation Report # F2008-34

Volunteer Fire Fighter Dies While Lost in Residential Structure Fire- Alabama

addition, NFPA 1002 *Standard for Fire Apparatus Driver/Operator Professional Qualifications* requires any driver/operator who will be responsible for operating a fire pump to also meet requirements of NFPA 1001 *Standard for Fire Fighter Professional Qualifications* for Fire Fighter I.[3, 19]

During this incident, fire fighters from the victim's department operated the tanker tasked with supplying water to fireground hoselines. The pump operator had received peer led hands-on training provided to him by members of this department. Fire fighters on scene recall having too much pressure on hoselines, not enough pressure, or seeing the pump panel unattended. The hoseline originally taken into the house by the victim was shut down when observed that it was flat. The closest section not exposed to fire was disconnected and another nozzle placed on it for use.

Recommendation #13: Fire departments should ensure that all fire fighters engaged in fireground activities wear the full array of personal protective equipment (PPE) issued to them.

Discussion: NFPA 1500 *Standard on Fire Department Occupational Safety and Health Program* contains the general recommendations for fire fighter protective clothing and protective equipment.[3] Chapter 7.1.1 specifies that "the fire department shall provide each member with protective clothing and protective equipment that is designed to provide protection from the hazards to which the member is likely to be exposed and is suitable for the tasks that the member is expected to perform." Chapter 7.1.2 states "protective clothing and protective equipment shall be used whenever the member is exposed or potentially exposed to the hazards for which it is provided." Chapter 7.2.1 states "members who engage in or are exposed to the hazards of structural fire fighting shall be provided with and shall use a protective ensemble that shall meet the applicable requirements of NFPA 1971 *Standard on Protective Ensembles for Structural Fire Fighting and Proximity Fire Fighting.*"[20]

On the day of incident, Charlie Fire Department members responded from work to the fire house to pick up CE1 and CE3. They responded to the incident scene without their department issued protective ensembles. Upon arrival to the incident, Alpha Fire Department members were trying to control the fire and make an attempt to enter the house for the victim. Their firefighting and rescue abilities were limited by the mutual aid department not having their gear with them.

Recommendation #14: Fire departments should ensure that fire fighters are trained to react to PASS and SCBA low air alarms, and that procedures are developed to properly shut down and secure a SCBA and its PASS device.

Discussion: During interviews, witnesses and fire fighters reported to NIOSH investigators that audible alarms were sounding from SCBAs doffed and placed on the fireground. NIOSH investigators are not sure if a PASS alarm on the victim could have been heard or distinguished from the alarms sounding outside. It was not determined if these alarms were from the lack of movement or a low air alarm. With alarms sounding and no one reacting to them, it is possible that search and

**Fatality Assessment and Control Evaluation
Investigation Report # F2008-34**

Volunteer Fire Fighter Dies While Lost in Residential Structure Fire- Alabama

rescue operations could be hindered or important radio messages not heard or understood, especially a brief Mayday radio call. During this incident, the victim did transmit a radio transmission requesting help early in the event that was heard by one fire fighter.

A fire fighter not reacting to the alarms is a symptom of desensitization (ignoring the sounds) and may cause a fire fighter to continue on with their assignment(s). If fire fighters are trained to react to alarms by investigating, then fire fighters can focus on the source of the alarm and take a positive action to correct it (i.e. responding to a downed fire fighter, properly securing the SCBA, or resetting the PASS device). Fire departments should develop standard operating procedures or guidelines that address proper doffing of a SCBA that is not in use, to include: turning the cylinder off, bleeding the line, and resetting or securing the PASS device.

Additionally,

Recommendation #15: States, municipalities, and authorities having jurisdiction should consider requiring mandatory training for fire fighters.

Discussion: Fire fighters have a high rate of injury death compared to other occupations.[21] Fire fighters need to take advantage of available training and certification programs within their jurisdiction so that safe and sound decisions can be made on the fireground. The state of Alabama has set mandatory minimal training requirements for an individual wishing to become a career fire fighter, but no mandatory or minimal training requirements have been set for individuals wanting to volunteer as a fire fighter.[2] Individuals may volunteer with a fire department and participate in fire ground activities in Alabama without being certified as a fire fighter. The state fire academy does offer a non-mandatory 160 hour certification course to individuals who are active volunteer members with a fire department and wish to become a certified volunteer fire fighter.[2] This course is equivalent to Fire Fighter I as outlined in NFPA 1001 *Standard for Fire Fighter Professional Qualifications* and can be completed over a 24 month period, but the victim and fellow fire department members had not taken the training. Since 2002, NIOSH has investigated three separate incidents that resulted in four fatalities of volunteer fire fighters in the state of Alabama.[22-24] NIOSH investigators cited the lack of training as a contributing factor in all of them.

REFERENCES

1. Alabama Fire College. So you want to be a volunteer fire fighter.
   [http://www.alabamafirecollege.org/wanttobeavolff.htm]. Date accessed: April 2009.

2. Alabama Fire College. Requirements for certified fire fighter.
   [http://www.alabamafirecollege.org/Commission/2007/360-X-2-2007.pdf]. Date accessed: April 2009.

Fatality Assessment and Control Evaluation Investigation Report # F2008-34

Volunteer Fire Fighter Dies While Lost in Residential Structure Fire- Alabama

3. NFPA [2008]. NFPA 1001 Standard for fire fighter professional qualifications. 2008 ed. Quincy, MA: National Fire Protection Association.

4. NFPA [2007]. NFPA 1500 Standard on fire department occupational safety and health program. 2007 ed. Quincy, MA: National Fire Protection Association.

5. NIOSH [1994]. NIOSH Alert: preventing injuries and deaths of fire fighters. Cincinnati, OH: U.S. Department of Health and Human Services, Centers for Disease Control and Prevention, National Institute for Occupational Safety and Health, DHHS (NIOSH) Publication No. 94-125. [http://www.cdc.gov/niosh/fire.html].

6. NFPA [2008]. NFPA 1561 Standard on emergency services incident management system. 2008 ed. Quincy, MA: National Fire Protection Association.

7. NFPA [2004]. NFPA 1720 Standard for the organization and deployment of fire suppression operations, emergency medical operations, and special operations to the public by volunteer fire departments. 2004 ed. Quincy, MA: National Fire Protection Association.

8. Endsley MR, Garland J [2000]. Situational awareness analysis and measurement. Mahwah, NJ: Lawrence Erlbaum Associates.

9. IFSTA [2008]. Essentials of fire fighting. 5th ed. Stillwater, OK: Fire Protection Publications, International Fire Service Training Association.

10. Occupational Safety and Health Administration (OSHA). Incident Command System eTool. [http://www.osha.gov/SLTC/etools/ics/index.html]. Date accessed: January 2009.

11. Carter W, Childress D, Coleman R, et al. [2000]. Firefighter's Handbook: essentials of firefighting and emergency response. Albany, NY: Delmar Thompson Learning.

12. FDSOA (Fire Department Safety Officers Association) [2002]. MAYDAY-MAYDAY-MAYDAY. By JJ Hoffman. Health and Safety for Fire and Emergency Service Personnel *13*(4):8.

13. Angulo RA, Clark BA, Auch S [2004]. You called mayday! Now what? Fire Engineering, *157*(9):93-95.

14. DiBernardo JP [2003]. A missing firefighter: give the mayday. Firehouse *Nov*:68-70.

15. Sendelbach TE [2004]. Managing the fireground mayday: the critical link to firefighter survival. [http://cms.firehouse.com/content/article/article.jsp?sectionId=10&id=10287]. Date accessed: February 2009.

**Fatality Assessment and Control Evaluation
Investigation Report # F2008-34**

Volunteer Fire Fighter Dies While Lost in Residential Structure Fire- Alabama

16. Miles J, Tobin J [2004]. Training notebook: mayday and urgent messages. Fire Engineering, *157*(4):22.

17. NFPA [2008]. NFPA 1521 Standard for fire department safety officer. 2008 ed. Quincy, MA: National Fire Protection Association.

18. IFSTA [1999]. Pumping apparatus driver/operator handbook. 1st ed. Stillwater, OK: Fire Protection Publications, International Fire Service Training Association.

19. NFPA [2009]. NFPA 1002 Standard for fire apparatus driver/operator professional qualifications. 2009 ed. Quincy, MA: National Fire Protection Association.

20. NFPA [2007]. NFPA 1971 Standard on protective ensembles for structural fire fighting and proximity fire fighting. 2007 ed. Quincy, MA: National Fire Protection Association.

21. Clarke C, Zak MJ [1999]. Fatalities to law enforcement officers and firefighters, 1992-1997. Compensation and Working Conditions.

22. NIOSH [2006]. Junior Volunteer Fire Fighter Dies and Three Volunteer Fire Fighters are Injured in a Tanker Crash – Alabama. Morgantown, WV: U.S. Department of Health and Human Services, Centers for Disease Control and Prevention, National Institute of Occupational Safety and Health, DHHS (NIOSH) Publication No. F2006-25. [http://www.cdc.gov/niosh/fire/reports/face200625.html].

23. NIOSH [2006]. Two Volunteer Fire Fighters Die When Struck by Exterior Wall Collapse at a Commercial Building Fire Overhaul - Alabama. Morgantown, WV: U.S. Department of Health and Human Services, Centers for Disease Control and Prevention, National Institute of Occupational Safety and Health, DHHS (NIOSH) Publication No. F2006-07. [http://www.cdc.gov/niosh/fire/reports/face200607.html].

24. NIOSH [2002]. Volunteer Fire Fighter Dies and Two are Injured in Engine Rollover – Alabama. Morgantown, WV: U.S. Department of Health and Human Services, Centers for Disease Control and Prevention, National Institute of Occupational Safety and Health, DHHS (NIOSH) Publication No. F2006-07. [http://www.cdc.gov/niosh/fire/pdfs/face200216.pdf].

INVESTIGATOR INFORMATION

This investigation was conducted by Stacy C. Wertman and Stephen T. Miles, Safety and Occupational Health Specialists with the Fire Fighter Fatality Investigation and Prevention Program, Fatality Investigations Team, Surveillance and Field Investigations Branch, Division of Safety Research, NIOSH located in Morgantown, WV. An expert technical review was provided by Harry R. Carter, Ph.D. The analysis of the victim's turnout gear was conducted by Jeff Stull, International

Fatality Assessment and Control Evaluation
Investigation Report # F2008-34

Volunteer Fire Fighter Dies While Lost in Residential Structure Fire- Alabama

Personnel Protection, Inc. Vance Kochenderfer, NIOSH Quality Assurance Specialist, National Personal Protective Technology Laboratory, conducted an evaluation of the victim's self-contained breathing apparatus.

NIOSH Fatality Assessment and Control Evaluation Investigation Report # F2008-34

Volunteer Fire Fighter Dies While Lost in Residential Structure Fire - Alabama

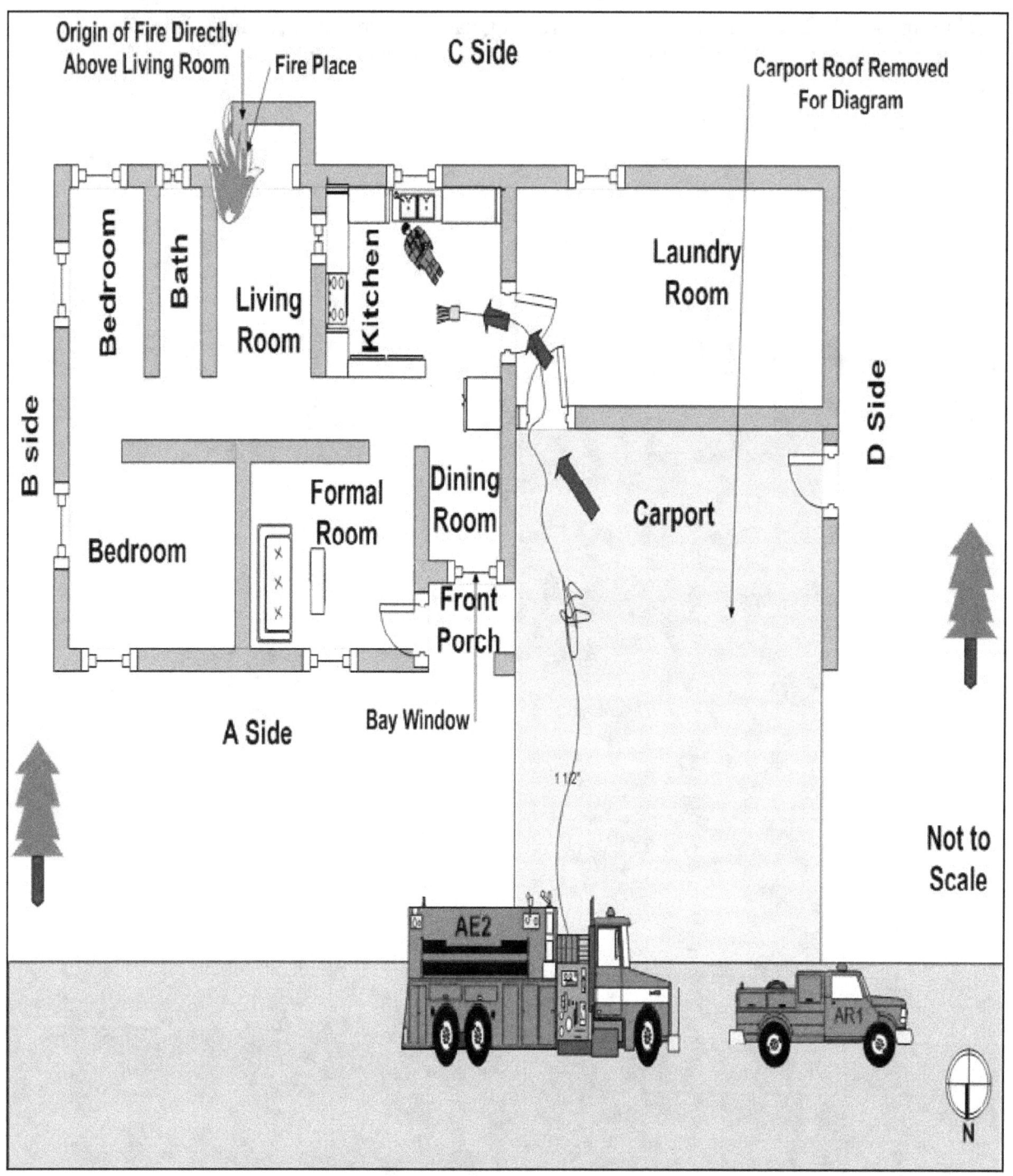

Diagram. Incident scene when victim called for help.

Death in the line of duty...

A summary of a NIOSH fire fighter fatality investigation
July 06, 2010

One Fire Fighter Killed and Eight Fire Fighters Injured in a Dumpster Explosion at a Foundry—Wisconsin

Executive Summary

In December 2009, a 33 year old male fire fighter died and eight fire fighters, including a lieutenant and a junior fire fighter, were injured in a dumpster explosion at a foundry in Wisconsin. At 1933 hours, dispatch reported a dumpster fire at a foundry in a rural area. Eight minutes later, the initial responding crews and the incident commander (IC) arrived on scene to find a dumpster emitting approximately two-foot high bluish green flames from the open top and having a ten-inch reddish-orange glow in the middle of the dumpster's south side near the bottom. The IC used an attic ladder to examine the contents of the dumpster: aluminum shavings, foundry floor sweepings, and a 55 gallon drum. Approximately 700 gallons of water was put on the fire with no affect. Approximately 100 gallons of foam solution, starting at 1 percent and increased to 3 percent, was then put on the fire, and again there was no noticeable effect. Just over twelve minutes on scene, the contents of the dumpster started sparking then exploded sending shrapnel and barrels into the air. The explosion killed one fire fighter and injured eight other fire fighters, all from the same volunteer department.

Contributing Factors

- Wet extinguishing agent applied to a combustible metal fire.
- Lack of hazardous materials awareness training.
- No documented site pre-plan.
- Insufficient scene size-up and risk assessment.
- Inadequate disposal/storage of materials.

A summary of a NIOSH fire fighter fatality investigation — Report #F2009-31

One Fire Fighter Killed and Eight Fire Fighters Injured in a Dumpster Explosion at a Foundry-Wisconsin

Key Recommendations

- Ensure that high risk sites such as foundries, mills, processing plants, etc. are pre-planned by conducting a walk through by all possible responding fire departments and that the plan is updated annually.

- Ensure that specialized training is acquired for high risk sites with unique hazards, such as combustible metals.

- Ensure that standard operating guidelines are developed, implemented and enforced.

- Ensure a proper scene size-up and risk assessment when responding to high risk occupancies such as foundries, mills, processing plants, etc.

- Ensure a documented junior fire fighter program that addresses junior fire fighters being outside the hazard zone.

Additionally, manufacturing facilities that use combustible metals should:

- Implement measures such as a limited access disposal site and container labeling to control risks to emergency responders from waste fires.

- Implement a bulk dry extinguishing agent storage and delivery system for the fire department.

- Establish a specially trained fire brigade.

The National Institute for Occupational Safety and Health (NIOSH) initiated the Fire Fighter Fatality Investigation and Prevention Program to examine deaths of fire fighters in the line of duty so that fire departments, fire fighters, fire service organizations, safety experts and researchers could learn from these incidents. The primary goal of these investigations is for NIOSH to make recommendations to prevent similar occurrences. These NIOSH investigations are intended to reduce or prevent future fire fighter deaths and are completely separate from the rulemaking, enforcement and inspection activities of any other federal or state agency. Under its program, NIOSH investigators interview persons with knowledge of the incident and review available records to develop a description of the conditions and circumstances leading to the deaths in order to provide a context for the agency's recommendations. The NIOSH summary of these conditions and circumstances in its reports is not intended as a legal statement of facts. This summary, as well as the conclusions and recommendations made by NIOSH, should not be used for the purpose of litigation or the adjudication of any claim.

For further information, visit the program Web site at www.cdc.gov/niosh/fire or call toll free 1-800-CDC-INFO (1-800-232-4636).

A summary of a NIOSH fire fighter fatality investigation   Report #F2009-31

One Fire Fighter Killed and Eight Fire Fighters Injured in a Dumpster Explosion at a Foundry-Wisconsin

Introduction

On December 29, 2009, a 33 year old male fire fighter died and eight fire fighters, including a lieutenant and a junior fire fighter, were injured in a dumpster explosion at a foundry in Wisconsin. On January 21 - 23, 2010, a general engineer from the National Institute for Occupational Safety and Health (NIOSH) Fire Fighter Fatality Investigation and Prevention Program conducted an opening meeting with the fire chief and conducted interviews with officers and fire fighters who were at the incident scene. The NIOSH investigator also visited the incident scene and met with the foundry's legal representatives, the Occupational Safety and Health Administration (OSHA) representative, the Wisconsin Department of Commerce's Fire Protection Coordinator, the Wisconsin Department of Natural Resources' Hazardous Waste Specialist, and the county sheriff's office representatives to review issues related to the site. The NIOSH investigator reviewed the officers' and fire fighters' training records, a video captured by the sheriff's patrol car of part of the incident scene, dispatch audio tapes, the county medical examiner's autopsy report and medical records of the injured fire fighters.

Fire Department

The volunteer department involved in this incident had 1 station with approximately 25 volunteer fire fighters and 7 fire apparatus serving a population of about 1,000 residents in a geographic area of approximately 25 square miles. The fire department averages 20 total calls per year.

The fire department had been called to this foundry a few times in the past for the smell of smoke in the building, but there were never any fires. The previous calls ended up being overheated drive belts on electrical motors. The fire department had no documented standard operating guidelines (SOGs).

Training and Experience

The table lists the training and experience of the primary fire fighters involved in the incident.

Fire Fighter	Injured (yes/no)	Training Courses	Years experience
Victim	yes - fatally	Fire Fighter 1, Basic Fire Investigation, Introductory ICS Level 100, Basic Lightweight Building Construction, Basic Wildland Fire Suppression, Managing Company Tactical Operations, and various other administrative and technical courses.	15
FF#1	yes	Various administrative and technical courses.	1
FF#2	yes	Entry Level Fire Fighter 1 and 2, Introduction to ICS 100 through 400, and various other administrative and technical courses.	8.5

A summary of a NIOSH fire fighter fatality investigation   Report #F2009-31

One Volunteer Fire Fighter Killed and Eight Fire Fighters Injured in a Dumpster Explosion at a Foundry—Wisconsin

FF#3	yes	Fire Fighter 1, various other administrative and technical courses.	7
FF#4	yes	Basic Fire Fighter 1, Advanced ICS, and various other administrative and technical courses.	8
FF#5	yes	Entry Level Fire Fighter 1 and 2, Introduction to ICS 100, Basic Wildland Fire Suppression, and various other administrative and technical courses.	6
FF#6	yes	On the Job Fire Fighter Training	3.5
Jr FF	yes	On the Job Fire Fighter Training	0.3
Lieutenant	yes	Fire Fighter 1, Incident Command System (ICS) Levels 100 through 400, Introduction to NIMS, The Dangers of Light weight Construction, and various other administrative and technical courses.	11
Captain	no	Fire Fighter 1, Entry Level Firefighter Part 1 and 2, Incident Command System (ICS) Levels 100 through 400 and 700, Incident Command, Critical Incident Management Response, Municipal Emergency Response Operations Part 1 and 2, Jaws of Life, and various other administrative and technical courses.	10
2nd Assistant Chief	no	Fire Fighter 1, Entry Level Firefighter Part 1 and 2, Entry Level Fire Officer, Incident Command System (ICS) Levels 100 through 400 and 700, Incident Command, Critical Incident Management Response, Pipeline Emergency Response, Jaws of Life, Wildland Fire Suppression, and various other administrative and technical courses.	9.5
1st Assistant Chief	no	Entry Level Firefighter Part 1 and 2, Entry Level Driver/Operator Part 1 and 2, Incident Command System (ICS) Levels 100 through 400 and 700, Incident Response to Terrorist Bombing Awareness, Recognizing and Identifying Hazardous Materials, Lightweight	29

A summary of a NIOSH fire fighter fatality investigation   Report #F2009-31

One Volunteer Fire Fighter Killed and Eight Fire Fighters Injured in a Dumpster Explosion at a Foundry—Wisconsin

			Construction Dangers, Jaws of Life, and various other administrative and technical courses.	
Chief (Incident Commander (IC))		no	Fire Fighter 1; Fire Officer I; Basic Fire Investigation; Introductory and Level 100, 200, and 300 ICS; Introduction to National Incident Management System (NIMS); HazMat I; Basic Lightweight Building Construction; Basic Wildland Fire Suppression; Managing Company Tactical Operations; and, various other administrative and technical courses.	31

Note: Fire Fighter 1 and 2 training met the criteria for National Fire Protection Association (NFPA) 1001, Standard for Fire Fighter Professional Qualifications, Fire Fighter I and Fire Fighter II.

Personal Protective Equipment

At the time of the incident, the victim and FF#4 were in full personal protective equipment (PPE) and had donned their self-contained breathing apparatus (SCBA) with integrated personal alert safety system (PASS). The other injured fire fighters, the lieutenant, and IC were wearing personal protective equipment consisting of turnout coat and pants, a helmet, and boots. Only the officers had portable radios.

Arrival Timeline of Apparatus and Personnel

1933 Hours
Dispatch reported a dumpster fire at a foundry in a rural area.

1941 Hours
Engine #11 (E11) - Chief (Driver and IC), captain, a fire fighter, and FF #1 (*injured*)
Tanker #17 (T17) – 1st Assistant Chief and FF #2 (*injured*)

1942 Hours
Engine #14 (E14) - Lieutenant (Lt) (*injured*), victim, FF #3 (*injured*), and FF #4 (*injured*)
Equipment Truck #15 (ET15) – 2nd Assistant Chief, driver and FF#6 (*injured*)

1945 Hours
Tanker #16 (T16) – FF #5 (*injured*)
Personally owned vehicle (POV) – Jr FF (*injured*)

A summary of a NIOSH fire fighter fatality investigation Report #F2009-31

One Volunteer Fire Fighter Killed and Eight Fire Fighters Injured in a Dumpster Explosion at a Foundry—Wisconsin

Notes: 1) The fire department is located approximately 5 miles from the foundry. 2) See Diagram 1 for placement of selected apparatus.

Weather

At the time of the incident, the weather conditions in the area were clear with an approximate temperature of 14 degrees Fahrenheit, 70 percent relative humidity, visibility 10 miles, and south-southwest winds at approximately 10 miles per hour.[1] Prior to the incident, a substantial amount of snow had fallen and the parking lot had been cleared which created a 4-foot snow bank around the parking lot perimeter.

Site Information

The incident occurred at a recycling dumpster on the grounds of a foundry which produced aluminum sand castings from various aluminum alloys. The 70,000 square foot foundry has been in business for over 73 years and currently employs over 100 people. The metal casting facility melts about 375,000 pounds of aluminum each month.

The fire department had walked through the facility 2 years prior to the incident, but there was no documentation and the hazards associated with the outside disposal area were not viewed or considered at that time. Several of the volunteer members had worked at the facility and/or have relatives that work there, thus they felt familiar with the facility.

The dumpster was made of 3/8 to 1/4 inch thick steel and measured approximately 17 feet long by 5 feet wide by 7 feet high. The dumpster contained about a month's worth of aluminum alloy shavings, foundry floor sweepings (consisting of dirt, metal particles, and processing fluids), and several open top 55 gallon drums of slag. *Note: Aluminum alloy slag is a by-product of the foundry's casting process and is produced during the separation of the molten aluminum alloy from impurities while in the aluminum casting furnaces. The slag occurs as a molten liquid melt and is a complex solution of silicates and oxides that solidifies upon cooling.* A representative dumpster and contents similar to the one that exploded are shown in Photo 1 and Photo 2. The dumpster was located in the southwest corner of the foundry's property. Dumpsters of different dimensions and contents were located in the same area, along with a utility pole, a commercial electrical panel box and a fenced-in electrical substation.

Within the previous 2 years, the foundry went from having 3 to 1 shifts per day, which reduced the amount of waste being produced. When the foundry was operating under 3 shifts per day, they kept the 55 gallon drums of slag stacked together and separate from the aluminum alloy shavings and the other metal scrap. After the reduction in shifts, the recycling company requested the foundry add the drums to the dumpster contents after the slag solidified, so the waste could be picked up in one load. Keeping the drums of slag separate from the open air dumpster contents per the previous storage method when 3 shifts were operating would ensure that no contaminates or heat transfer from the slag could mix with the dumpster contents.

A summary of a NIOSH fire fighter fatality investigation Report #F2009-31

One Volunteer Fire Fighter Killed and Eight Fire Fighters Injured in a Dumpster Explosion at a Foundry—Wisconsin

Investigation

The following investigation synopsis details events leading up to the fatal incident. A deputy sheriff patrol car video recording of the south-side of the incident scene and interviews from the on-scene fire fighters were used to reconstruct the events leading to the incident. The victim was part of a second crew pulling a 1 ¾" hoseline into position when the explosion occurred.

On December 29, 2009, a 33 year old male fire fighter died and eight fire fighters including a lieutenant and a junior fire fighter less than 18 years of age were injured in a dumpster explosion at a foundry in Wisconsin. At 1933 hours, dispatch reported a dumpster fire at a foundry. At 1941 hours, Engine 11 (E11) with the Fire Chief (IC) driving and Tanker 17 (T17) arrived on scene to find a dumpster emitting two-foot high bluish green flames from the open top and having a ten inch reddish-orange glow in the middle of the dumpster's south side near the bottom. The E11 captain manned the pumper while FF#1 and another fire fighter pulled a 1 ¾" hoseline to the west side of the dumpster. The IC called for FF#2 to get the attic ladder from E11 so he could examine the contents of the dumpster. After stepping on the first rung, the IC noticed aluminum shavings, foundry floor sweepings, and a 55 gallon drum in the dumpster. After observing the contents and color of the flames, the IC believed that metal cutting fluids and/or oils were burning. Two foundry employees were on scene and reassured the IC that no magnesium was in the dumpster.

At 1942 hours, Engine 14 (E14) and Equipment Truck 15 (ET15) arrived on scene. E14 pulled up to the south of E11 and ET15 staged in the east side parking lot. The IC then requested the 1st assistant chief and another fire fighter to set up the dump tank. With a charged hoseline, positioned 20 feet away from the west-side of the dumpster and standing up on a 4-foot snow bank, FF#2 began to flow water using the fog nozzle, and was backed up by FF#1 and another fire fighter. The lieutenant (LT) from E14 took over the fog nozzle after telling FF#2 to put on an air pack. The hoseline crew flowed approximately 700 gallons of water at 50 pounds per square inch (psi) with no affect on the fire, which prompted the IC to call for foam.

At 1945 hours, Tanker 16 (T16) arrived on scene followed by a Jr FF in a privately owned vehicle (POV). The Jr FF reported to the 1st assistant chief and was instructed to go to E14 to get a radio to listen for tool requests. T17 left the scene to refill. The victim and FF#4 were pulling another 1 ¾" hoseline from E11 while FF#2 and FF#3 straightened out the hoseline (see Diagram 1). FF#6 assisted T16 in backing up to the dump tank and was preparing to assist with the tender dumping.

About 100 gallons of foam solution (approximately 6 gallons of Class A/B foam concentrate) from E11's internal foam induction system, starting at 1 percent and increased to 3 percent at 100 psi, was put on the fire and again there was no noticeable effect. At approximately 1953 hours, just as the IC was calling to discontinue foam operations, the contents of the dumpster started sparking, then within seconds, exploded sending shrapnel and barrels into the air. The explosion fatally injured the victim and injured eight other fire fighters (see Diagram 2, Diagram 3, Photo 3, Photo 4, and Photo 5).

A summary of a NIOSH fire fighter fatality investigation   Report #F2009-31

One Volunteer Fire Fighter Killed and Eight Fire Fighters Injured in a Dumpster Explosion at a Foundry—Wisconsin

Fire Behavior

Key characteristics of this fire were:

- a metal fire (primarily aluminum alloy shavings) with approximately 2-foot high bluish green flames and a 10 inch reddish-orange glow at the bottom of the steel dumpster;

- approximately 700 gallons of water was put on the fire with no change in fire intensity and the water turning into steam;

- about 100 gallons of Class A/B foam solution was put on the fire with again no change in fire intensity;

- then white sparks began shooting into the air and there was an explosion.

Per the state fire marshal's report, the cause and origin of the fire is listed as undetermined and the cause of the explosion was a result of the fire suppression efforts and the introduction of water and suppressant foam.[2]

After conducting a combustible metal literature review, one speculative theory is that a thermite reaction started from aluminum shavings and particles mixed with metal oxides or silicon oxides (wet sand) which generated enough energy to ignite the aluminum shavings and particles. (*Note: A thermite reaction is a pyrotechnic composition of a metal powder and a metal oxide, which produces an exothermic chemical reaction using aluminum as the reducing agent at high temperature. Thermites can be a diverse class of compositions. The fuels are often aluminium, magnesium, calcium, titanium, zinc, silicon, and boron. The oxidizers can be boron(III) oxide, silicon(IV) oxide, chromium(III) oxide, manganese(IV) oxide, iron(III) oxide, iron(II,III) oxide, copper(II) oxide, and lead(II,III,IV) oxide. The most common thermite is aluminium-iron(III) oxide.*)[3] Once started, the thermite reaction does not need air from the outside to continue burning. The addition of wet extinguishing agent (in this case, water and a foam solution) on the fire most likely generated hydrogen gas, due to the volatile reaction with the aluminum, which exploded.

Factors that may have contributed to the thermite reaction are: the last slag barrel put in the dumpster was still too hot to touch (normally allowed to cool 24 hours but in this case only 1 ½ to 3 hours – initial temperature of the slag (aluminum oxide) is 1425 degrees Fahrenheit when put inside barrel); and the iron (III) oxide (commonly known as rust) in the dumpster and on the slag barrels. *Note: Thermal pattern and iron oxide on slag barrels shown in photo 2.* The melting of the snow laying on top of the dumpster contents may have initially started the generation and/or release of hydrogen gas.

The reaction of small aluminum shavings/particles with any or all of the above mentioned factors has the potential to cause a thermite reaction. Reportedly, the slag barrel was too hot to touch and only cooled for 1 ½ to 3 hours when placed in the dumpster. The temperature of the slag barrel could have been in the 700 to 800 degree Fahrenheit range and may have directly initiated the thermite reaction. In addition to silicon oxides being part of the alloy process, the iron (III) oxide or rust on the slag barrels and on the walls of the steel dumpster provided another means for the most common of

A summary of a NIOSH fire fighter fatality investigation Report #F2009-31

One Volunteer Fire Fighter Killed and Eight Fire Fighters Injured in a Dumpster Explosion at a Foundry—Wisconsin

thermite reactions (aluminum-iron (III) oxide) to occur. The thawing or melting snow could have further hydrated the iron oxide and/or silicon oxide which could have enhanced the energy being released when mixed with the aluminum shavings/particles to cause the ignition of the aluminum shavings.

The National Fire Protection Association (NFPA) 484, *Standard for Combustible Metals*, Annex A Explanatory Material, paragraph A.13.3.3.10.3, states that the application of a wet extinguishing agent (particularly water hose streams) accelerates a combustible metal fire and could result in an explosion. In addition, paragraph A.13.3.3.10.1, states water reacting with aluminum can give off highly flammable hydrogen gas.[4,5]

Contributing Factors

Occupational injuries and fatalities are often the result of one or more contributing factors or key events in a larger sequence of events that ultimately result in the injuries or fatality. The NIOSH investigator identified the following items as key contributing factors in this incident that ultimately led to the line of duty death of one fire fighter and to the injuries of eight fire fighters:

- Wet extinguishing agent applied to a combustible metal fire.
- Lack of hazardous materials awareness training.
- No documented site pre-plan.
- Insufficient scene size-up and risk assessment
- Inadequate disposal/storage of materials.

Cause of Death/Injuries

According to the medical examiner's autopsy report, the victim died from multiple injuries as a result of blunt force trauma. According to medical records, all of the fire fighters' injuries were due to the explosion resulting in debris impact and/or noise-related injuries. Of the fire fighters injured at the time of the incident, the Lt had lower back and spinal cord injuries; FF #1, FF#5, FF#6 and the Jr FF experienced temporary hearing loss; FF#2 had back pain and temporary hearing loss; FF#3 had a neck sprain and a bump on his head; FF#4 had second degree burns to the left elbow, right flank, and a broken right hand.

Recommendations

Recommendation #1: Fire departments should ensure that high risk sites such as foundries, mills, processing plants, etc. are pre-planned by conducting a walk through by all possible responding fire departments and that the plan is updated annually.

Discussion: National Fire Protection Association (NFPA) 1620 *Standard for Pre-Incident Planning, 2010 Edition*, states that the pre-incident plan should be the foundation for decision making during an emergency situation and provides important data that will assist the IC in developing appropriate strategies and tactics for managing the incident. This standard also states that the primary purpose of a

A summary of a NIOSH fire fighter fatality investigation Report #F2009-31

One Volunteer Fire Fighter Killed and Eight Fire Fighters Injured in a Dumpster Explosion at a Foundry—Wisconsin

pre-incident plan is to help responding personnel effectively manage emergencies with available resources. Pre-incident planning involves evaluating the protection systems, building construction, contents, and operating procedures that can impact emergency operations. Section 8.1 states "The pre-incident plan shall identify and document any special hazards recognized by the authority having jurisdiction that present extraordinary life safety challenges, operations challenges, or other challenges to emergency responders."[6] A pre-incident plan identifies deviations from normal operations and can be complex and formal, or simply a notation about a particular problem such as the presence of flammable liquids, explosive hazards, common attic, drop ceilings, roofing materials, modifications to structural building components, or structural damage from a previous fire.[7,8]

In addition, NFPA 1620 outlines the steps involved in developing, maintaining, and using a pre-incident plan by breaking the incident down into pre-, during- and post-incident phases. In the pre-incident phase, for example, it covers factors such as physical elements and site considerations, occupant considerations, protection systems and water supplies, hydrant locations, and special hazard considerations. The pre-incident plan should be documented, shared with other departments who provide mutual aid, and if possible, entered into the dispatcher's computer so that the information is readily available if an incident is reported at the noted address.

In this incident, the fire department had walked through the facility 2 years prior to the incident but there was no documentation and the hazards associated with the outside disposal area were not indentified or considered at that time. Several of the fire fighters had worked at the facility and/or have relatives that work there, thus they felt familiar with the facility.

Recommendation #2: Fire departments should ensure that specialized training is acquired for high risk sites with unique hazards, such as combustible metals.

Discussion: Fire departments often respond to complex or unique hazards which require specialized/advanced knowledge and/or training in dealing with that hazard. Combustible metal fires present unique and dangerous hazards to fire fighters which are not commonly encountered in conventional fire fighting operations. The temperatures encountered in a combustible metal fire far exceed those of a structure fire.[5]

The National Fire Protection Association (NFPA) 484, *Standard for Combustible Metals*, states that it is extremely important to conduct a good size-up by identifying the combustible metals involved, the physical state of the metals (e.g., shavings, chips, fine dust, etc.), the location relative to other combustible materials, and the quantity of the product involved. NFPA 484, A.13.3.3.10.3, states that the application of a wet extinguishing agent (particularly water hose streams) accelerates a combustible metal fire and could result in an explosion.[4] This is due to the water reacting with aluminum to give off highly flammable hydrogen gas. This conversion of water into hydrogen has a heat value (British Thermal Units per pound (Btu/lb)) of about 2.8 times that of gasoline, assuming 100 percent conversion of the hydrogen in the water. This equates to flowing 42.8 gallons per minute (gpm) of gasoline on the fire for every 100 gpm of water. Thus, in lieu of using a wet extinguishing agent, primarily water, it is recommended that a bulk dry extinguishing agent be used such as dry sand, dry soda ash, or dry sodium chloride. If no bulk dry agents are available, the best approach may be to

A summary of a NIOSH fire fighter fatality investigation   Report #F2009-31

One Volunteer Fire Fighter Killed and Eight Fire Fighters Injured in a Dumpster Explosion at a Foundry—Wisconsin

isolate the material as much as possible, protect exposures, and allow the fire to burn out naturally.[5] Proper training is a must to properly identify and handle these unique fires. Manufacturers and fire departments with combustible metals in their jurisdiction should review Chapter 13 of the National Fire Protection Association (NFPA) 484: Standard on Combustible Metals.[4]

Recommendation #3: Fire departments should ensure that standard operating guidelines are developed, implemented and enforced.

Discussion: Written SOGs enable individual fire department members an opportunity to read and maintain a level of assumed understanding of operational procedures. Conversely, fire departments can suffer when there is an absence of well developed SOGs. The NIOSH Alert, *Preventing Injuries and Deaths of Fire Fighters* identifies the need to establish and follow fire fighting policies and procedures.[9] Guidelines and procedures should be developed based on recognized standards and best practices, trained upon by all potential responding personnel, and fully implemented and enforced to be effective. The following NFPA Standards identify the need for written documentation to guide fire fighting operations:

NFPA 1500 *Fire Department Occupational Safety and Health Program* states that fire departments shall prepare and maintain policies and standard operating procedures that document the organizational structure, membership, roles and responsibilities, expected functions, and training requirements, including the following....(4) The procedures that will be employed to initiate and manage operations at the scene of an emergency incident. In particular, NFPA 1500, 4.2.3.1, states a risk management plan should include risk identification of actual and potential hazards.[10]

NFPA 1561 *Standard on Emergency Services Incident Management System* states that standard operating procedures (SOPs) shall include the requirements for implementation of the incident management system and shall describe the options available for application according to the needs of each particular situation.[11]

Recommendation #4: Fire departments should ensure a proper scene size-up and risk assessment when responding to high risk occupancies such as foundries, mills, processing plants, etc.

Discussion: Per NFPA 1620, the foundry would be classified as an *Industrial Occupancy* and is defined as "an occupancy in which products are manufactured or in which processing, assembling, mixing, packaging, finishing, decorating, or repair operations are conducted.[6] When responding to a high-risk occupancy such as a foundry, the first arriving officer or incident commander must conduct a proper scene size-up of the incident including a risk assessment in terms of life safety for responders and occupants. Size-up is defined as an ongoing process of evaluating the situation to determine what has happened, what is happening, and what is likely to happen. A sufficient scene size-up should include:

- ✓ Type of Occupancy
- ✓ Access

A summary of a NIOSH fire fighter fatality investigation Report #F2009-31

One Volunteer Fire Fighter Killed and Eight Fire Fighters Injured in a Dumpster Explosion at a Foundry—Wisconsin

- ✓ Building Construction
- ✓ Environmental Conditions
- ✓ Location of the Fire
- ✓ Resources Responding
- ✓ Water Supply
- ✓ Special Hazards/Risks
- ✓ Time of Day
- ✓ Color of Smoke
- ✓ Utilities
- ✓ Built-in Fire Protection
- ✓ Pre-Incident Planning

Based upon the type of occupancy and what is burning, the IC needs to contact and establish communications with the occupant/facility representative regarding the situation. A relationship with the occupant/facility representative should initially take place during a pre-incident planning visit. NFPA 1, *Fire Code* states that an occupant/facility shall designate and train a liaison representative for the fire department. A liaison shall assist in pre-incident planning and emergency response procedures which identify the location of hazardous materials.[12] If there is no life safety hazard (danger to civilians/occupants) and the fire does not extend or threaten other parts of the facility, the IC should stage all resources until an incident action plan is developed. Once what is burning is properly identified, the appropriate extinguishing agent should be used. If the responding fire departments do not have the resources needed to safely attack the fire (e.g. a bulk dry extinguishing agent for combustible metal fires) and the fire is not threatening lives, the contents should just be allowed to burn out while protecting exposures.

NFPA 1500 states in Chapter 8 – Emergency Operations, that risk management principles must be routinely employed by supervisory personnel at all levels of the incident management system to define the limits of acceptable and unacceptable positions and functions for all members at the incident scene. The risk to fire department members is the most important factor considered by the Incident Commander in determining the strategy that will be employed in each situation.[10]

Recommendation #5: *Fire departments should ensure a documented junior fire fighter program that addresses junior fire fighters being outside the hazard zone.*

Discussion: The involvement of junior fire fighters (teenagers less than 18 years of age) in the fire service was established early in the fire service's history. Particularly, volunteer departments have welcomed teenage personnel, often alongside of other family members serving in the department. A positive lifelong connection can be achieved when starting young people out within a fire department. However, being that they are under 18 years of age creates certain responsibilities for the fire department.

Many state and federal laws govern work-related activities of youths, which protect the health/safety and educational opportunities of young people. The Fair Labor Standards Act (FLSA), the primary law addressing child labor, exempts volunteer activities, however, the U.S. Department of Labor

A summary of a NIOSH fire fighter fatality investigation   Report #F2009-31

One Volunteer Fire Fighter Killed and Eight Fire Fighters Injured in a Dumpster Explosion at a Foundry—Wisconsin

advocates voluntary compliance with FLSA requirements for work-like volunteer activities. In addition, OSHA (Federal) has jurisdiction over volunteers in an employer-employee relationship.[13] In this case, OSHA had no jurisdiction because the employer-employee relationship did not exist.

A fire department establishing a program for junior fire fighters should document what the allowable duties are for junior fire fighters in a department's SOGs. The National Volunteer Fire Council (NVFC) has published a *Junior Firefighter Program Handbook* that addresses the steps of starting and maintaining a program, including sample documents from currently existing programs for reference.[14] The International Association of Fire Chiefs (IAFC) has published *Opening New Doors: Guidelines and Best Practices for a Successful Youth Fire Service Program*.[15] If a junior fire fighter would have been near the same location of the victim, death or severe injury could have resulted. Thus, at no time should a junior fire fighter ever be potentially in harms way.

In this case, the junior fire fighter was aware of his duties and was not directly involved in fire suppression. Due to the limited area around the incident, the hazard zone encompassed where the apparatus were staged. However the junior fighter could have been more safely utilized at the station assisting in the tanker refilling, away from the hazard zone. The temporary hearing loss incurred was a result of the limited area around the incident and the unexpected intensity of an explosion.

Recommendation #6: Fire departments should **ensure** *all fire fighters who may operate in or near a hazard zone, prior to approaching, have donned the full complement of personal protective equipment, i.e., self-contained breathing apparatus and turn-out gear*.

Discussion: Although there is no evidence that the use of personal protective equipment (PPE) would have prevented this fatality or these injuries, this recommendation is provided as a good safety practice. Properties of burning metals cover a wide range. Even when the exact metal is known, metal fires should never be approached without proper protective equipment. Additionally, PPE may not protect wearers from injuries due to contact with molten materials as this hazard exceeds design criteria of PPE and wearers must be aware of PPE limitations. The toxicity of certain metals is also an important factor in fire suppression. Some metals (particularly heavy metals) can be fatal if they enter the bloodstream or their smoke fumes are inhaled.[16]

At any dumpster fire, fire fighters should be fully protected and on-air prior to approach; approach can be at a distance where cooling and control can start well before approaching the dumpster. Fire fighters should remember that there is nothing worth saving in a dumpster fire as the contents are always unknown. The dumpster may contain class A combustibles, discarded hazardous materials, or propane tanks. A dumpster fire is literally a potential bomb and should be treated in that way.

In this incident, only two fire fighters on scene were wearing their full complement of PPE.

A summary of a NIOSH fire fighter fatality investigation — Report #F2009-31

One Volunteer Fire Fighter Killed and Eight Fire Fighters Injured in a Dumpster Explosion at a Foundry—Wisconsin

Recommendation #7: Manufacturing facilities that use combustible metals should implement measures such as a limited access disposal site and container labeling to control risks to emergency responders from waste fires.

Discussion: Manufacturing facilities should have a standard operating procedure (SOP) for the disposal of their manufacturing waste. This SOP should include adequate labeling of containers and storage areas for these wastes. It should also identify unique hazards if the public were to come in contact with the waste or if fire fighters or other emergency responders were to respond to an emergency. NFPA 704, *Identification of the Hazards of Materials for Emergency Response*, states all buildings or areas storing, using, or handling hazardous materials be marked by use of a standardized placarding system. The placarding system identifies hazard categories for health, flammability, reactivity and special hazards, including water reactivity and oxidizers.[17] This SOP should be reviewed annually for any changes in the materials being stored for waste. In addition, to help prevent uncontrolled dumping of unknown materials into the dumpster, a secured area may be necessary.

Within the previous 2 years, the foundry went from having 3 to 1 shifts per day, which reduced the amount of waste being produced. When the foundry was operating under 3 shifts per day, they kept the 55 gallon drums of slag stacked together and separate from the aluminum alloy shavings and the other metal scrap. After the reduction in shifts, the recycling company requested the foundry add the drums to the dumpster after the slag solidified so the waste could be picked up in one load. A safer procedure might be to keep the drums of slag separate from the open air dumpster contents per the previous storage method when 3 shifts were operating. NFPA 484, section 13.2.7.3.2, states that open storage of metal chips and dust particles that are readily ignitable should be isolated and segregated from other combustible materials and metal scrap to prevent propagation of a fire.[4]

In this incident, there was no labeling on the dumpster that identified the contents. The Wisconsin Department of Natural Resources determined that no hazardous waste regulations were violated.

Recommendation #8: Manufacturing facilities that use combustible metals should implement a bulk dry extinguishing agent storage and delivery system for the fire department.

Discussion: Fire departments do not routinely have bulk dry extinguishing agents on hand or a method to implement their use. The National Fire Protection Association (NFPA) 484, *Standard for Combustible Metals*, Annex A Explanatory Material, paragraph A.13.3.3.10.3, states that facilities that use combustible metals should provide these dry extinguishing agents in areas near where the combustible metals are used and stored, including scrap. These bulk dry extinguishing agents must be kept free of moisture wherever they are stored. A delivery system to apply the dry extinguishing agents should be onsite, such as a front end loader and/or dump truck, and it should be available to the fire department at all times. *Note: Automatic sprinkler protection systems for areas where combustible metals are used or stored are not recommended.*[4]

A summary of a NIOSH fire fighter fatality investigation Report #F2009-31

One Volunteer Fire Fighter Killed and Eight Fire Fighters Injured in a Dumpster Explosion at a Foundry—Wisconsin

Recommendation #9: Manufacturing facilities that use combustible metals should establish a specially trained fire brigade.

Discussion: A specially trained fire brigade that is knowledgeable in the permissible methods of fighting incipient fires within the areas that utilize or contain the combustible metals can be a very effective means of fire protection. Only specially trained personnel should be permitted to engage in fire control activities; all other personnel should be evacuated. Upon arrival of the local fire department, all fire control activities should be by a unified incident command that includes knowledgeable plant personnel.[4]

References

1. Weather Station History: Weather Underground [2009]. http://www.wunderground.com/weatherstation/WXDailyHistory.asp?ID=KWISHERW1&month=12&day=29&year=2009 Date accessed: March 2009.

2. Wisconsin Division of Criminal Investigation [2010]. Case Report Number: 09-5094.

3. Wikipedia Foundation, Inc. [2010]. http://en.wikipedia.org/wiki/Thermite Date accessed: April 2010.

4. NFPA [2009]. NFPA 484: Standard on Combustible Metals. Quincy, MA: National Fire Protection Association.

5. Fire Engineering [2008]. Proper Handling of Combustible Metal Fires. http://www.fireengineering.com/index/articles/display/320694/articles/fire-engineering/volume-161/issue-2/features/proper-handling-of-combustible-metal-fires.html Date accessed: March 2010.

6. NFPA [2010]. NFPA 1620: Standard for pre-incident planning. Quincy, MA: National Fire Protection Association.

7. Klaene BJ, Sanders RE [2000]. Structural fire fighting. Quincy, MA: National Fire Protection Association.

8. International Fire Service Training Association [2008]. Essentials of fire fighting and fire department operations, 5th ed. Stillwater, OK: Fire Protection Publications, Oklahoma State University.

9. NIOSH [1994]. NIOSH alert: preventing injuries and deaths of fire fighters. Cincinnati, OH: U.S. Department of Health and Human Services, Public Health Service, Centers for Disease Control and prevention, National Institute for Occupational Safety and Health, DHHS (NIOSH) Publication No. 94-125

A summary of a NIOSH fire fighter fatality investigation

Report #F2009-31

One Volunteer Fire Fighter Killed and Eight Fire Fighters Injured in a Dumpster Explosion at a Foundry—Wisconsin

10. NFPA [2007]. NFPA 1500: Standard on fire department occupational safety and health program. Quincy, MA: National Fire Protection Association.

11. NFPA [2008]. NFPA 1561 Standard on emergency services incident management system. Quincy, MA: National Fire Protection Association.

12. NFPA [2009]. NFPA 1: Fire Code. Quincy, MA: National Fire Protection Association.

13. Varone, Curt. "How Old Is Old Enough? Legal Considerations for Junior Firefighter Program." Firehouse Magazine February 2010: Fire Law. Print

14. National Volunteer Fire Council [2009]. Junior Firefighter Program Handbook. http://juniors.nvfc.org/cms/images/stories/resources/department/juniorhandbook.pdf Date accessed: April 2010

15. International Association of Fire Chiefs [2010]. Opening New Doors: Guidelines and Best Practices for a Successful Youth Fire Service Program. http://www.iafc.org/displayindustryarticle.cfm?articlenbr=41787 Date accessed: May 2010

16. DOE [1994]. Department of Energy Handbook, Primer on Spontaneous Heating and Pyrophoricity. http://www.hss.energy.gov/NuclearSafety/ns/techstds/standard/hdbk1081/hbk1081b.html Date accessed: March 2010.

17. NFPA [2007]. NFPA 704: Identification of the Hazards of Materials for Emergency Response. Quincy, MA: National Fire Protection Association.

Investigator Information

This incident was investigated by Matt Bowyer, General Engineer, with the NIOSH Fire Fighter Fatality Investigation and Prevention Program, Division of Safety Research. The report was authored by Matt Bowyer. Expert technical reviews were conducted by Deputy Chief William Goldfeder, Loveland-Symmes Fire Department and editor of FirefighterCloseCalls.com and by Murrey E. Loflin, Director, West Virginia University Fire Service Extension. A technical review was also provided by the National Fire Protection Association, Public Fire Protection Division. An external chemical opinion on thermite reaction was given by Professor Alan Stolzenberg, Department of Chemistry, West Virginia University. Additional external comments were provided by the Occupational Safety and Health Administration's local area office and the foundry's legal representative and their task force manager.

A summary of a NIOSH fire fighter fatality investigation   Report #F2009-31

One Volunteer Fire Fighter Killed and Eight Fire Fighters Injured in a Dumpster Explosion at a Foundry—Wisconsin

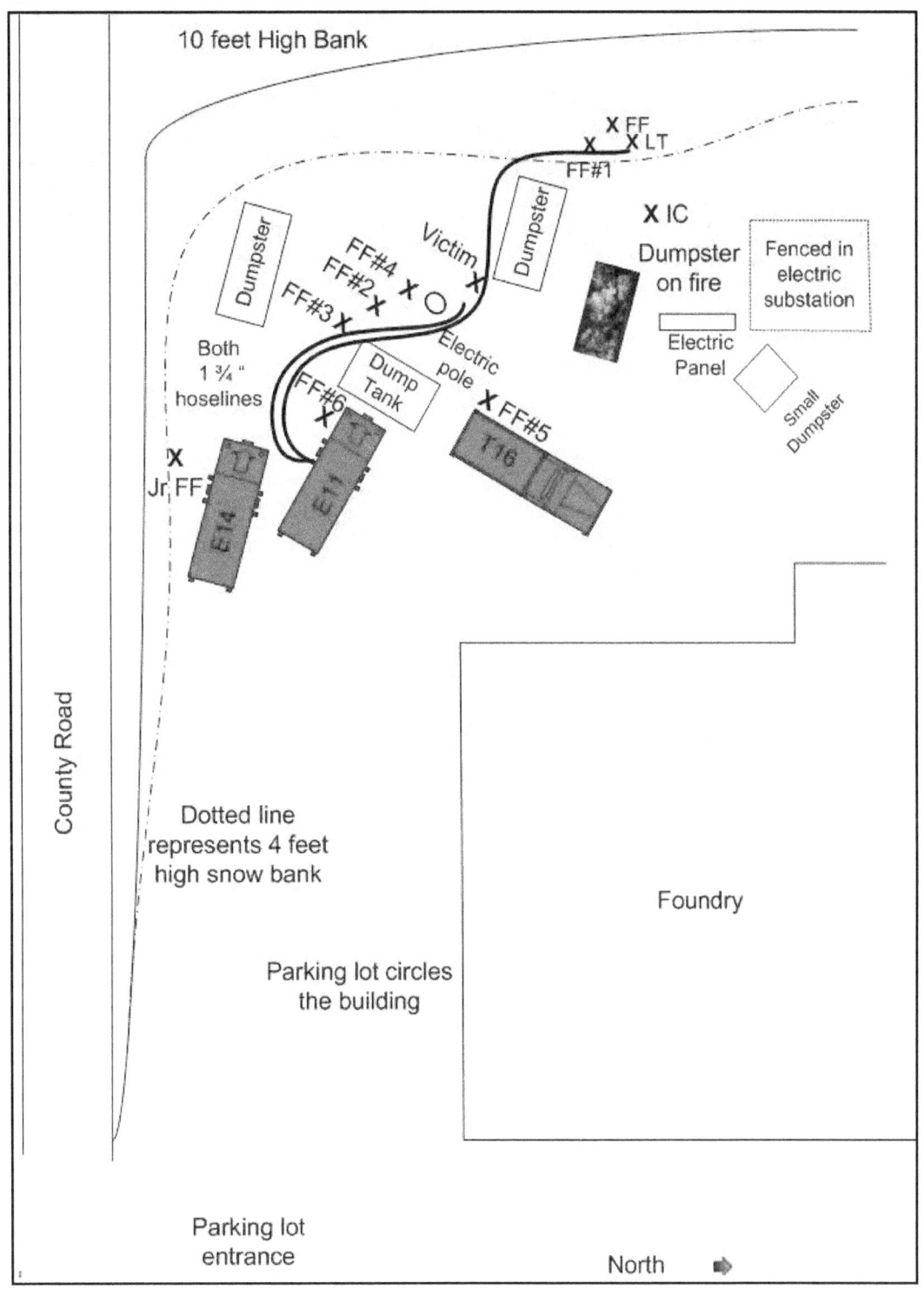

Diagram 1. Approximate locations of key apparatus and personnel just prior to the explosion.

A summary of a NIOSH fire fighter fatality investigation

Report #F2009-31

One Volunteer Fire Fighter Killed and Eight Fire Fighters Injured in a Dumpster Explosion at a Foundry—Wisconsin

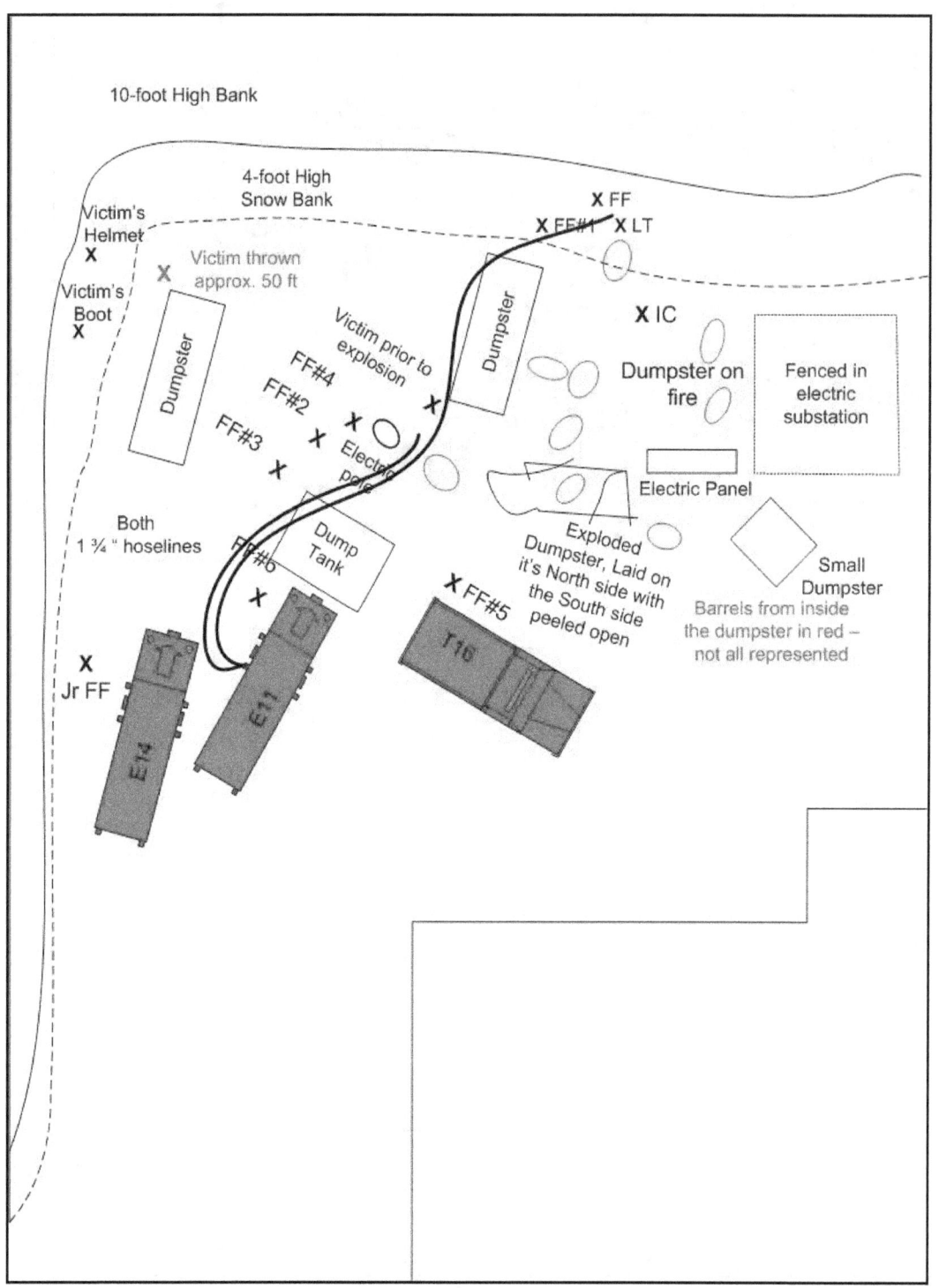

Diagram 2. Approximate location of dumpster and victim after the explosion.

A summary of a NIOSH fire fighter fatality investigation

Report #F2009-31

One Volunteer Fire Fighter Killed and Eight Fire Fighters Injured in a Dumpster Explosion at a Foundry—Wisconsin

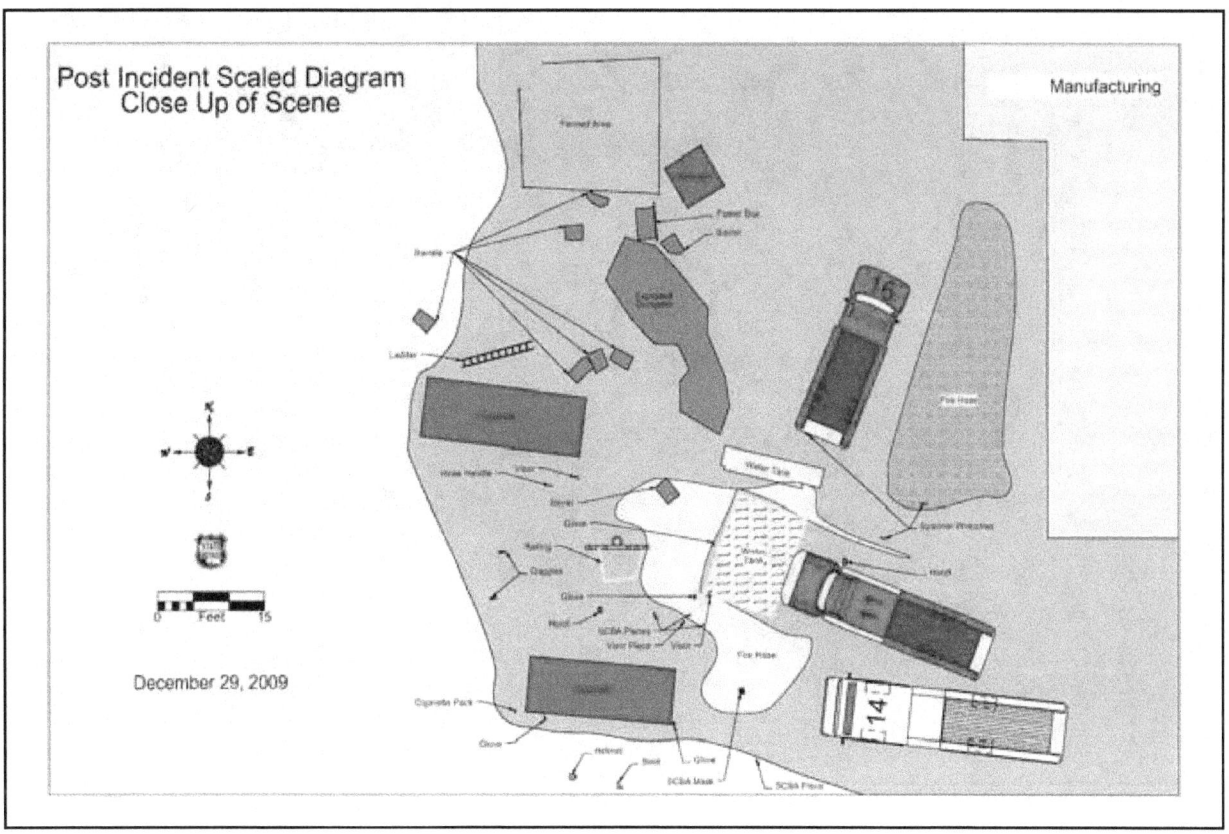

Diagram 3. Post incident scaled close-up of the scene.
(*Courtesy of the Sheriff's Department*)

A summary of a NIOSH fire fighter fatality investigation Report #F2009-31

One Volunteer Fire Fighter Killed and Eight Fire Fighters Injured in a Dumpster Explosion at a Foundry—Wisconsin

Photo 1. A dumpster similar in size and shape to the one that exploded. (*NIOSH photo*)

A summary of a NIOSH fire fighter fatality investigation

Report #F2009-31

One Volunteer Fire Fighter Killed and Eight Fire Fighters Injured in a Dumpster Explosion at a Foundry—Wisconsin

Photo 2. Dumpster contents similar to the contents of the dumpster that exploded. Note thermal pattern and iron oxide (rust) on slag barrels.
(*NIOSH photo*)

A summary of a NIOSH fire fighter fatality investigation Report #F2009-31

One Volunteer Fire Fighter Killed and Eight Fire Fighters Injured in a Dumpster Explosion at a Foundry—Wisconsin

Photo 3. Barrel that was blown from the dumpster and believed to have impacted the lieutenant who had been standing at that spot on the snow bank with the 1 ¾" hoseline.
(*NIOSH photo*)

A summary of a NIOSH fire fighter fatality investigation Report #F2009-31

One Volunteer Fire Fighter Killed and Eight Fire Fighters Injured in a Dumpster Explosion at a Foundry—Wisconsin

Photo 4. Exploded dumpster and its' remaining contents. (*NIOSH photo*)

A summary of a NIOSH fire fighter fatality investigation   Report #F2009-31

One Volunteer Fire Fighter Killed and Eight Fire Fighters Injured in a Dumpster Explosion at a Foundry—Wisconsin

Photo 5. Bottom of the dumpster showing the origin of the fire in the dumpster.

(*NIOSH photo*)

www.ingramcontent.com/pod-product-compliance
Lightning Source LLC
Chambersburg PA
CBHW081830170526
45167CB00007B/2777